GARDEN OF INVENTION

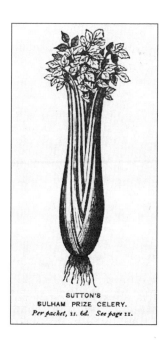

SUTTON'S
SULHAM PRIZE CELERY.
Per packet, 1s. 6d. See page 11.

Dedicated
to the
Pioneers of
Gardening

GARDEN OF INVENTION

The Stories of Garden Inventors & Their Innovations

George Drower

THE LYONS PRESS
GUILFORD, CONNECTICUT
An imprint of The Globe Pequot Press

First Lyons Press revised edition, 2003.

Originally published in English by Sutton Publishing under the title *Gardeners, Gurus & Grubs* copyright © George Drower 2001.

The Lyons Press is an imprint of The Globe Pequot Press.

10 9 8 7 6 5 4 3 2 1

Printed in the United States of America

ISBN 1-58574-779-3

Library of Congress Cataloging-in-Publication data is available on file.

Contents

Introduction

Everyone has heard of the celebrated garden "designers" such as "Capability" Brown, Humphrey Repton, and Gertrude Jekyll, who transformed the gardens of grand country estates. But what of the less famous inventors and innovators? Today their seemingly simple contraptions and ideas are often taken for granted and they themselves are largely forgotten. Yet these were the *real* heroes of gardening. Not only did they provide many of the components that made possible the innovations of the more famous designers, but their ingenious contraptions went on to improve the quality of ordinary gardens.

In unearthing these neglected pioneers, the circumstances that brought about their inventions, and the improvements their ideas made, this book makes some astonishing discoveries. It seems that in every garden, no matter how basic, there will be at least a handful of the simplest garden components—from flowerpots to Gro-Bags, wheelbarrows to tree-movers, flame-guns to bamboo canes—whose stories are fascinating and varied. So who were these unsung heroes and how did their ideas develop?

There was the Governor of Brittany, for example, the Marquis Bertrand de Moleville, who fled across the English

Channel in fear of losing his head during the French Revolution. Thankful to have escaped the guillotine he spent many of his long years in exile in England pondering how to design an effective pair of garden cutters for beheading roses. The result was the revolutionary secateurs—now an indispensable gadget for most gardeners—which eventually went into production in 1818. In France they were an immediate success with fruit farmers and rose growers, being safer to use than conventional billhooks and pruning knives.

Another émigré in London during those years was the Marquis de Chabannes, whose family's estates near Bordeaux had been seized by the great unwashed proletariat. He set up in business near Covent Garden as a "posh plumber"! Effectively he "adapted" the ideas of Dr. Bonnemain, a brilliant yet obscure Parisian physician, who had devised a stove for incubating hens' eggs by means of circulating hot water through pipes. With this in mind Chabannes created a system for heating glasshouses by means of hot water pipes (and by seeking to apply that idea to residential properties developed the first modern form of hot water domestic central heating).

Credit for designing the world's first effective greenhouse belongs to Salomon de Caus, a clever mathematician from Dieppe who was engaged in restructuring Heidelberg Castle for the Elector Palatine in the early seventeenth century. In 1619 he built a massive, 240-foot-long prefabricated structure, with internal heating and large windows, to protect the Elector's orange trees during the winter months. While at Heidelberg de Caus also devised an ingenious machine for raising water for garden fountains by means of steam. He apparently went insane because no one would acknowledge him as the inventor of the world's first steam pump, and he died in a Parisian lunatic asylum where his invention was, it seems, copied by Lord Worcester—who later claimed it as his own.

Other gardening pioneers never received due recognition for their endeavors. Captain Thomas Hardwicke had discovered, in Nepal, the distinctive *Rhododendron arboreum*, although he seldom received acknowledgment for introducing it to England. Instead, all the credit went to Nathaniel Wallich, whose Himalayan rhododendron flourished—the plants having been transported to England in tins of sugar! Another inventor who missed out was Sir Christopher Cockerell. His hovercraft, though a technical success, was a commercial flop as a passenger-carrier, whereas Denmark's Karl Dahlman, who adapted the same simple hovercraft principles to devise the hover-mower, created a phenomenally successful garden product with the Flymo.

Some of the most far-reaching inventions and innovations were brought about by altruists like Nathaniel Ward, a doctor in London's East End. In 1832, while attempting to breed butterflies in a glass jar, Ward happened to notice that a fern seedling, accidentally included in the butterfly jar, was growing sturdily, regardless of its enclosed condition. Effectively Ward had discovered the principle of micro-climate. He went on to create a range of closely glazed cabinets in which plants were kept moist by their own circulating vapor. At the Great Exhibition in 1851 he displayed a cabinet containing perfectly healthy plants which had not required watering for eighteen years! These "Wardian cases" revolutionized the international transportation of plants, enabling delicate species to be taken on long voyages without ill effects. Indeed, by this means James Hooker conveyed thousands of tea plants from China to India, and thereby founded the Indian tea industry. Also known as "terrariums," the Wardian cases now enabled the affluent middle classes to grow plants in their living rooms, and—much to Ward's delight—provided the urban poor with an affordable means of including in their lives an element of greenery which might improve their physical and psychological well-being.

Then in 1915 came a remarkable discovery. An army surgeon, Colonel Charles Cathcart, a big-hearted yet unassuming man who had played rugby for Scotland and had already devised several ingenious medical contraptions, found that, when treated, sphagnum moss could be utilized as an ideal form of bandage for treating gunshot wounds. Cathcart organized public collections of the moss, and in the process proved that moss was worthy of respectability and was not just a nuisance on lawns or a lining for hanging baskets. Doubtless many military lives were saved by this doctor and his moss bandages.

There were also entrepreneurs, whose garden gadgetry inventions had far-reaching consequences. Henry Bewley, for example, developed an alternative to rubber for making garden hoses. Gutta percha, plus the machine by which he extruded it, was subsequently found to be ideal for use as insulation for the earliest transatlantic telephone cables. Then there was the enterprising Telford family, whose simple idea of a priced catalog started the development of the practice of buying garden products by mail-order.

Another businessman was William Barron, whose creation of a heavy-duty vertical tree transplanter enabled vast cedars of Lebanon and mature yews (some as tall as 39 feet) to be shuffled about the countryside, making Barron the first proponent of the "instant garden." At the rather less glamorous end of the gardening spectrum is the commercially produced Gro-Bag, which was unveiled in 1973 and has since allowed millions of green-fingered enthusiasts to experience the pleasures of horticulture, even if only on a balcony.

Many of the most useful innovations are derived from the ancient world. Chuko Liang, a Chinese general, is credited with being the originator of the wheelbarrow; Theophrastus, Aristotle's deputy in the third century B.C., devised the earliest plant classification; then, some four centuries later, the notion of the modern-style lawn was invented by Pliny the Younger, while the silver-tongued Roman statesman Cicero conferred respectability on the humble trellis.

Surprisingly, even several centuries ago, many garden pioneers were concerned about the damage being done to the environment. Columella, the Roman agriculturist whose simple advice on soil testing remains a useful guide for gardeners today, argued that soil fertility could be maintained by proper manuring. In the Bahamas the eighteenth-century illustrator Mark Catesby was horrified at the obliteration of entire forests of native trees. Meanwhile, in Pennsylvania America's most famous nurseryman and plant gatherer John Bartram expressed concern that fossils were being removed (although he himself had dispatched a few choice specimens across the Atlantic). In Suffolk Eve Balfour was so concerned about the quantity of chemicals being used on agricultural and horticultural land that she founded the Soil Association, which soon provided the ground rules for the modern organic movement.

Despite having the best of intentions a few garden pioneers actually *damaged* the environment. John Bennet Lawes's superphosphate was effectively the world's first artificial fertilizer; introduced in the early 1840s, it was advertised in *The Gardeners' Chronicle* as a great boon for horticulturists.

However, the process of manufacturing the magical compound involved vast quantities of sulphuric acid—which had a serious impact on the health of people living near Bennet Lawes's Thamesside factories. Most spectacular of the gardening *faux pas* has been the infamously fast-growing Leyland hedge tree. Christopher Leyland, the philanthropic millionaire who discovered it in 1888, would doubtless have been distraught to learn that his splendid Leyland tree would cause so many neighborly disputes.

Rarely was the quintessential image of an inventor more inappropriate than in the case of these garden pioneers. Such individuals were far from the stereotypical characters of disheveled, flustered appearance, with holes in their pullovers, laboriously developing their eccentric ideas in a workshop. In fact, of the pioneers covered in the many stories in this book surprisingly few were professional inventors, or even career gardeners. Many had drawn their ideas from related fields such as agriculture, commerce, adventuring, water technology, and steam engineering. A quite amazingly high proportion were involved with medicine, as altruistic bush doctors or as qualified physicians and surgeons. Yet although the garden pioneers were almost invariably astonishingly important people in those other areas, they all shared a common delight in, and love of, gardens.

Tools of the Trade

Chuko Liang's Wheelbarrow

The simplest and oldest of tools are invariably the most useful, and none more so than the wheelbarrow. Its inventor was Chuko Liang (181–234 A.D.), sometimes known as Zhuge Liang, who was prime minister of Shu and the author of *Three Kingdoms*, a classic Chinese work on warfare and political strategy. One of the warring "Three Kingdoms" at the time of the Han Empire, Shu was a vast state located in western China. A great and ingenious general, the technically minded Chuko Liang grappled with the logistical problem of how to supply his armies in muddy soil conditions and hilly terrain, especially when there was a severe labor shortage. With some assistance from Li Zhuan, a naturalist and engineer with whom he had been improving the crossbow, Chuko Liang had the first wheelbarrows constructed. (It has been suggested that relief carvings on the Wu Liang tomb-shrines near Shandong, dated to 147 A.D., show a vehicle which *could* be a wheelbarrow, but the carvings are faded and the interpretation uncertain.)

The newfangled prototype that trundled out of a workshop

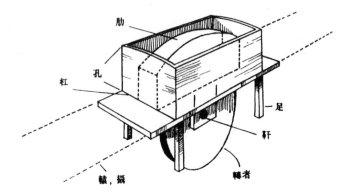

肋

孔

杠

足

軒

轅，輈

輀者

A reconstruction of Chuko Liang's original army service wheelbarrow. (Reproduced by permission of Cambridge University Press from Needham, *Science and Civilisation in China*)

in 231 A.D. to shift army supplies at a place called Jiangzhou was indeed a remarkable contraption. The wheel, which was very large in proportion to the barrow, was placed in the center of the part on which the load was laid, so that all the weight bore upon the axle. The barrowman supported no part of the weight, but served merely to move the barrow forward, and keep it in equilibrium. The wheel was cased up in a frame made of laths, and covered over with a plank 4 or 5 inches wide. On each side of the barrow was a projection, on which the goods were balanced—rather like the saddle of a pack-horse. The wheel was at least 3 feet in diameter, with spokes which were slight and numerous. Its rim, which at first glance appeared to be unsuitably slender, was deliberately narrow: in the rainy season it would cut through boggy ground in which broader-rimmed wheels would stick fast. From this basic design Chuko Liang soon devised two simpler variants: the "Wooden Ox," so called because its shafts projected in front and it was pulled; and then the "Gliding Horse" wheelbarrow, which was pushed.

The use of a large wheel enabled Chinese wheelbarrows to cope with rough terrain and they were used for moving goods over long distances. (Reproduced by permission of Cambridge University Press from Needham, *Science and Civilisation in China*)

Chuko Liang's invention apparently reached western Europe via the Byzantine Empire where it was encountered by westerners during the Second Crusade. They took the idea back home and modified their existing handcarts to ride on one wheel only. The earliest known western depiction of a wheelbarrow appears in a stained-glass window at Chartres Cathedral, dated 1220 A.D. Initially wheelbarrows were much used for traveling long distances, such as moving goods across Alpine passes. However, later they were normally confined to short-haul work. Unlike the see-saw Chinese models, the medieval European types had the wheel placed well forward and were low-slung. Although more stable they had smaller cargo capacity.

Much of the garden wheelbarrow's evolution took place on medieval building sites. Although the wheelbarrow was six times more expensive to buy than a two-wheeled handcart, only one laborer was required to push it. Effectively it could pay for itself within a week. And because one operative could now shift the load of two, the wheelbarrow was an absolute must for working gardeners. Numerous barrow shapes developed. Gardeners had a choice of wooden or wrought-iron wheelbarrows. Some had sides, others a framework for holding pots, and yet others only a stopper-board in front. Then there was

Variations of the wheelbarrow's basic design were experimented with throughout the world. In the early 1860s the *American Gardener's Monthly* reported a sighting of this ingenious makeshift watering barrow. (*Journal of Horticulture and Cottage Gardener*, August 9, 1864)

pebbles. For hothouses, greenhouses, and structures in which flat paving or tiles is used as flooring, a broom of whalebone or bast, inserted in small bundles, in a rectan-

gular piece of wood, and fastened in with wire after the manner of a brush, is most suitable. Inferior brooms of this sort are made by putting the bast into the holes made for its reception, and fastening it in with hot pitch;

FIG. 307.—WOODEN WHEELBARROW.

these may do good service for a short time, but they are by no means durable, and therefore are not cheap. Brooms of iron or copper wire are sometimes used for

mossy lawns, paths overgrown with moss, and for clearing moss from the trunks of trees, but they are seldom if ever seen now. A birch broom costs from 3d. to 6d., and a good bast broom from 1s. 6d. to 2s.

633. **Appliances for Carriage and Transfer.** — These are requisite to a greater or less degree in every garden, according

FIG. 308.—IRON WHEELBARROW.

to its size and the work that is to be done in it. First, and most important of all is the wheelbarrow, used for the conveyance of mould, manure, weeds, litter, etc., from

one part of the garden to the other, and from the stable yard and manure heap to the garden and *vice versâ*. Next in importance to the wheelbarrow is the handbarrow, of which there are different forms, calculated to serve different purposes. Then

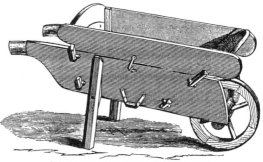

FIG. 309.—WOOD BARROW WITH MOVABLE TOP.

come smaller articles, serving for conveyance of earth, manure, pots, flowers in pots, etc., and weeds and litter, from place to place, among which are the mould box or

By 1885 Mrs. Beeton's *Book of Garden Management* demonstrated the many different designs commercially available to gardeners.

the "haulm barrow" (haulm meaning discarded plant stems) for carrying litter, leaves, and prunings. Many of these low-slung models had a light carrying capacity, such as a basket for grass-cuttings. Also devised was the "separating barrow," the body of which was secured by bolts, enabling it to be lifted off so that the load could be taken into hothouses where a whole barrow could not go. Chuko Liang's brilliant invention—in its European form—appeared to have reached its apotheosis in the eighteenth century when a tree-carrying barrow was created.

Yet on the other side of the world, in the Chinese coastal province of Shantung, a visiting Dutch diplomat beheld an amazing innovation: sail-assisted wheelbarrows! In 1794 an astonished van Braam Houckgeest watched junk-rigged barrows moving overland. Rapidly sketching some illustrations in his journal he noted: "Each of them had a sail, mounted on a small mast in a socket arranged at the forward end of the barrow. The slat sail, made of matting, or more often of cloth, is five or six feet high, and three or four feet broad, with stays, sheets, and halyards, just as on a Chinese boat. The sheets join the shafts of the wheelbarrow and can thus be manipulated by the barrowman in charge." Just

A "sail-barrow." The extent to which the Chinese had evolved the wheelbarrow into a high-tech machine astonished the visiting Dutch diplomat van Braam Houckgeest when he saw Chinese laborers using junk-rigged wheelbarrows on dry land in the 1790s. (Reproduced by permission of Cambridge University Press from Needham, *Science and Civilisation in China*)

how many such "sail-barrows" were used in large gardens is unknown. Surely Chuko Liang could never have envisioned that his simple invention would become so technically advanced.

For a detailed history of Chuko Liang's era, check the useful websites www.ageofkings.org and www.inventors.about.com.

Bertrand de Moleville's Secateurs

The man who invented every gardener's most useful gadget, the secateurs, seldom if ever receives a mention, but on those rare occasions when he does, his name is almost certainly given as M. Bertrand of Moleville. In fact his full name was Antoine-François Bertrand de Moleville, and the initial "M" stands not for a first name or Monsieur, but for Marquis. Why the Marquis de Moleville should have sought to invent secateurs is really quite a mystery, but he certainly made the invention in exciting circumstances.

Although de Moleville was a fairly prominent politician, surprisingly little is known about his personal life. An ardent supporter of King Louis XVI, de Moleville was Governor, and effectively the military commander, of Brittany. His monarchist stance made him a much-loathed figure, as he discovered in 1788 when he ventured forth from the fortress at Rennes to announce an edict—and was pelted with stones by a furious mob, who even threw a noose around his neck. Fleeing France in fear of his life, in 1789 he became an émigré in Britain. In London he busied himself writing to a senior statesman, the Duke of Portland, offering to inform him of the best places for British forces to attack the Brittany coast. He wrote a book, *Private Memoirs Relative to the Last Year of Louis XVI* (1793), in

Before de Moleville's time, secateurs tended to be more decorative than practical, such as these sixteenth-century ornamental overhead branch-trimmers. (*The Gardeners' Chronicle*, 1877)

support of the late king, then another in 1803 vituperatively condemning the author Helen Maria Williams for daring to criticize that royal dictator.

At the end of the Napoleonic period it seems de Moleville returned to France and, despite his earlier treasonable readiness to facilitate a British invasion, for a while held a ministerial post in the French government. Out of office again, this time he occupied himself inventing the secateurs, although by what means he produced and developed his prototype remains unknown. However, there was certainly a need to create such a gadget. For centuries the standard tools for pruning and vine-trimming had been the knife and billhook, both dangerous to use, especially if dropped from a height. De Moleville's new "secateurs"—the French word for cutters—had two curved blades. The edge of each was beveled in opposite directions, so that the flat blades worked smoothly across each other. The blades were fastened together by a rivet, around which they turned. When not in use they were held together by a strap at the ends of the handles; when open, the blades were forced apart and held in that position by a spring between the handles.

By 1815 de Moleville had completed the basic design work, creating the famous cutters which were introduced to the public three years later. Initially he seems to have expected them to be used primarily by vineyard workers, although in 1818 the annual directory *Le Bon Jardinier* referred to them as the latest invention which might replace the pruning knife, so the makers were evidently already targeting them at gardeners. The timing was perfect. Secateurs proved particularly useful for the new rose hybrids. These, unlike their shrubby predecessors, required careful pruning to ensure blooming in subsequent years. William Robinson noted in his *Gleanings from French Gardens:* "A secateur is seen in the hands of every French fruit-grower, and by its means he

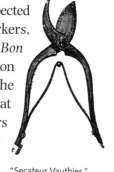

"Secateur Vauthier," which was similar to de Moleville's original design for the common secateurs. (Robinson, *Gleanings from French Gardens*, 1868)

cuts as clean as the best knife-man with the best knife ever whetted. They cut stakes with them also as fast as one could count them." Nevertheless in those early years the secateurs were not universally popular. They had many detractors, especially in Britain where for a long time gardeners regarded them as merely a woman's accouterment. (Indeed gardening magazines advertised them as such.)

Soon there were various types of secateurs, with differing forms of blades, springs, and handles. The most popular were the Lecointe, with a coiled spring; the Vauthier, with a notch for cutting wire (1864); the Aubert, with a single spring (1865); and then the anvil secateur.

The Marquis de Moleville escaped the guillotine in the French Revolution, and

A later variant was the "Secateur Lecointe," with a coiled spring. William Robinson, an avid francophile, was influential in encouraging British gardeners to use secateurs rather than a pruning knife.

died knowing that he had established a means by which countless gardeners could in the future protect their hands while pruning.

 A range of early secateurs and clippers can be seen on display at the Museum of Garden History, Lambeth Palace Road, London SE1 7LB; see also their website at www.compulink.co.uk/~museumgh.

George Acland's Jute Twine

Beloved for its organic simplicity, an indispensable tool for every gardener's pocket is a ball of twine, which—if it is green—will almost certainly be made of jute. It is taken for granted now, yet until 1828 there was no jute twine in Europe. Previously, gardeners who wanted to tie plants to a frame, or string up their runners, had to use hazel strips or raffia, which was of unreliable strength and not readily available in urban areas.

The founder of the jute manufacturing industry was Englishman George Acland, who began his nautical career as a midshipman in the Royal Navy and then served with the East India Marine Service. Upon leaving the Service he took up commercial activities, first in Ceylon and later in Bengal. Acland realized the commercial potential of jute when he saw it being used by Oriya gardeners employed in the botanical gardens of the East India Company. They called it "jhut," a term from which the modern name seems to have derived. At this time, jute was still virtually unknown outside India where its fibers had been used for centuries to make twine, cord, and coarse fibers. It grew best in a hot, moist atmosphere in areas with considerable rainfall, and most was produced in India's Bengal province, where it flourished, especially in the highland districts. The stalks, which were either cut down with a sickle or pulled up by hand, were gathered into bundles and immersed in stagnant pools or streams to undergo the process known as "retting," which loosened the fibers and separated them from the stem. To speed up this process, the operator would stand in the pool and beat and shake the stems to strip away the resinous bark. After the fibers had been agitated in the water to remove the vegetable impurities, they were wrung out and suspended on a line to dry in the sun. The fibers would then be made up into bundles and carefully sorted according to quality.

Yet a means had never been devised of producing it by a mechanical process. Acland got in touch with manufacturers of paper at Serampore who were experimenting with fibers in the hope of improving and cheapening their output, and this seems to have prompted the idea in Acland's mind of finding a better means of making twine. He became excited— although the market for cotton, hemp, flax, and other fibers was huge, they were relatively expensive. Jute was not as strong, but it was hard-wearing and tear-resistant, and had the advantage of being much cheaper. Thus it would be the ideal material for garden jobs for which the requirements were medium strength and low cost.

Acland went to Scotland to raise capital for his new business venture and to find a location for a factory. He opted for Dundee, which at that period was an important textile center based around flax and hemp; it was, consequently, only natural that the longer, coarser, but otherwise apparently similar jute fibers should be tested on the machinery already used for the preparation and spinning of flax and hemp. Or so Acland thought. In 1828 he established the Chapelshade works, where he struggled to overcome the many difficulties that resulted from the use of unsuitable machinery, because jute is far more woody and brittle than either flax or hemp. These difficulties, however, were gradually overcome. It was found that by mixing the fibers of two plants, *Corchorus olitorius* and *Corchorus capsularis*, an effective substitute for flax could be produced. Although the best jute was inferior in durability and strength to hemp and flax, and even single strands were of variable tenacity throughout their length, it was found that by making the twine three-ply its strength could be greatly increased.

Manufacturing began in 1832, yet business was initially slow. But in 1838 a representative of the Dutch government placed a large order with Acland for jute bags to be used for carrying the crop of coffee beans from West Africa. More than any other factor, that fortunate event led to the rapid growth of the industry. While Dundee was developing into an international center for the manufacture of garden twine it was suggested to Acland that he should send the machinery to Bengal because the jute could be more economically spun there. From that suggestion came the building of the first mill in Calcutta in 1854. Ironically it was also the first step in Dundee's decline, and eventual collapse, as a jute products center.

In common with most other textile fibers, jute is susceptible to degradation from mildew, sunlight, and heat; it can also be damaged by water, insects, and rodents. Realizing that the useful life of the fiber could be prolonged by appropriate chemical treatment, Acland began soaking his garden twines in copper compounds and green creosote. Thus by the time he died Acland had not only made twine affordable for gardeners, he had also established its distinctive traditional green appearance.

Visit the Ropery at the World Naval Base, Historic Dockyard, Chatham, Kent; www.medwaytown.com/dockyard.

William Barron's Tree-Mover

The legendary transplanting of the Domesday Book's "Buckland Yew" near Dover in 1880 was a sensational feat of tree-moving. It was the crowning achievement of William Barron, the nineteenth century's greatest expert at shifting huge trees; among his other accomplishments he had secured a name for himself as the creator of the first stately home garden open to the paying public, and was a pioneer of the "instant garden" concept.

A distinctive means of moving trees had been devised by "Capability" Brown, who for the purposes of landscaping Stowe in the 1740s had produced a two-wheeled machine, called a "yanker" or "timber wheels." The wheels supported a long beam or pole which would be strapped to the trunk and hauled horizontally, thereby with a fork-like action yanking the hapless tree from the ground. When creating the fabulous gardens at Versailles André Lenôtre had used a similarly cumbersome method for extracting trees—as a consequence of which many died. Sir Henry Steuart, who in 1827 wrote *The Planter's Guide*, took things a stage further by insisting that trees be prepared for moving a year or two in advance by having trenches dug around them and then filled in to encourage the growth of a mass of fibrous roots before the

Lancelot Brown's "yanker" machine was chained to a tree and then pulled down to the horizontal, thereby literally tearing the tree from the ground. (Loudon, *An Encyclopaedia of Gardening*, 1822)

The disadvantage of this sort of machine was that the tree's rootball could be damaged while the tree was being moved. (*The Planter's Guide*, 1827)

move took place. Despite this, Steuart was a leading advocate of the Brownite yanking technique; although he recognized there could be advantages in hoisting trees vertically, he also favored pulling trees down. Once down, the trees were transported on oak sleds, and on reaching the new planting site each tree would be raised into its new position by means of horses heaving a rope over crossed poles. However, it was William Barron who developed a reliable means of moving really large trees.

Sir Henry Steuart used an even harsher method of moving trees: dragging them on sleds then hauling them upright by means of a horse heaving a rope over upright poles. (*The Gardener's Magazine*, June 1828)

Born at Eccles in Berwickshire on September 7, 1800, William Barron was from an old Scottish family which originated in the neighborhood of Aberdeen. William proved to be a bright boy at school, where he studied Greek and French, and learned sufficient Hebrew to enable him to read the Bible in the original. Having developed a taste for horticulture, William was apprenticed to Lady Honstan Boswell's garden at Blackadder, Berwickshire. During his apprenticeship he studied his profession with such effort that he obtained a post at the Botanic Garden, Edinburgh; here, much to his surprise, he was put in full charge of the greenhouses. He continued working there for three years, during which time he doubtless gained some experience of moving light shrubs on the two-wheeled contraption designed by the head gardener at Edinburgh, William McNab. This lightweight apparatus was probably created for the purpose of shifting small trees when the Botanic Garden moved to a new site in Edinburgh in 1824. Having expressed a wish to specialize in pine forcing, Barron was sent by McNab to the Duke of Northumberland's estate at Syon House, Isleworth. While there, he assisted in planting its large conservatory. Then, in March 1830, Barron was appointed to oversee the Earl of Harrington's gardens at Elvaston Castle, Derbyshire, where virtually nothing had ever been done to improve or ornament the grounds.

Although he did have something of an artistic mind, William delighted in tinkering with mechanical contraptions. Soon after his arrival at Elvaston that August he compulsively set about repairing a double-pistoned water-engine he had found abandoned there. That winter there was civil unrest and Nottingham Castle was burned down by a mob, who subsequently turned their attention to Elvaston. Though the

William Barron. (*The Gardeners' Chronicle*, 1891)

castle was attacked and a serious fire broke out, Barron's repaired engine helped to bring the blaze under control. Upon arriving to inspect the damage, Harrington's first words were "Barron, you have saved my castle." William told him: "I have only done my duty." The episode had the effect of proving Barron's considerable powers of ingenuity and, unintentionally, securing the absolute trust of his employer. By then Barron's character was well formed. A man of indomitable perseverance, he had profound religious faith and was one of the pioneers of the Temperance cause at a time when such a position was difficult to maintain.

In 1831 the fourth Earl of Harrington married the beautiful though promiscuous actress Maria Foote (1798–1867), previously implicated in several salaciously high-profile affairs. This time the union was a love match, and for the next twenty years, having retreated from Society, the earl and his actress lived at Elvaston in total solitude. The earl's instructions were: "If Queen Victoria comes, show her around, but admit no one else." Harrington had asked "Capability" Brown to lay out the gardens there, but Brown refused, arguing that the land was too flat. So the work was carried out by William Barron instead. It made his name and enabled him to go on to rival Brown.

The terrain at Elvaston was indeed flat and low, as was the adjoining country, being only a few feet above the level of the River Derwent which sluggishly flowed close by. Unperturbed, Barron spent the next four years busily draining the ground and establishing a kitchen garden. In November 1830, Harrington declared that he wanted to move three cedar of Lebanon trees, each 39 feet tall. Barron duly dug trenches around the trees to encourage new roots, and then, having filled them in, he prepared to wait a year. But the earl insisted that he wanted them moved in February—just four months later! Barron had been unimpressed by Steuart's written work and knew enough about it to realize that the yanker machines would never be able to shift such big trees. However, on seeing Harrington's disappointment, he told him: "No such trees have ever been successfully removed before, but if your lordship will support me I will form a plan of my own and remove your trees and make them grow."

A disadvantage of the transplanting machines used by Henry Steuart was that the trees suffered from having their roots laid bare. Barron thought that trees should be planted with their roots encased in a ball of earth and that instead of being dropped into a hole the rootball should be rested on the ground and a new mound built up around it. In order to move trees with their heavy rootballs intact, Barron designed a sturdy four-wheeled cart which could raise and lower them vertically on chains by means of lever-operated ratcheted windlasses. Having built a suitably strong machine, in February 1831 he successfully moved all three cedars of Lebanon to the East Avenue. Then in November he shifted another cedar, this one 52 feet high. Clearly this was a means by which dramatic effects could instantly be brought to the unpropitiously flat terrain at Elvaston.

From this time until the death of the fourth earl, tree-moving on the most extensive scale was practiced at Elvaston. Barron scoured the country to find specimens, and many old yew trees, some of them hundreds of years old, were brought

Tree-lifting machine: side elevation. (Robinson, *Gleanings from French Gardens*, 1868)

distances of 19 miles or more. The surrounding countryside became like a scene from H. G. Wells's *War of the Worlds*, with

villagers agog to see Barron's machines lurching along the rutted lanes bearing cargoes of towering trees. Thousands of native specimens, and even already-pruned topiary, were transported by this means.

The pleasure grounds at Elvaston, which were practically nonexistent in 1831, were gradually extended and most

A heavy tree-lifting machine, similar to that devised by William Barron: end view. (Robinson, *Gleanings from French Gardens*, 1868)

elaborately laid out. By 1851 the fire-damaged castle had been repaired and the garden sported an arboretum containing virtually every known species of conifer. Major avenues were lined with parallel rows of Irish and golden yews, and even monkey puzzle trees. The centerpiece of the entire 580-acre garden, which was based on a sixteenth-century design, was a topiary collection. Reputedly, the garden also contained some 11 miles of yew and other evergreen hedges. Not only had Barron begun and completed a garden that was one of the most distinctive and regal in Britain, but he had done it virtually "instantly."

On inheriting Elvaston in 1851 the fifth earl found himself in somewhat awkward financial circumstances, resulting from his father's lavish expenditure on the estate, inspired by the love of an actress. The impoverished successor, who needed to cut back on the maintenance of the gardens to help make ends meet, took the drastic step of opening Elvaston to the general public. It is usually thought that the first stately home to be commercially opened to the public was Longleat in 1949 (although in 1776 the Earl of Pembroke had let a select few members of the public into his Wilton House free of charge). At Elvaston the threadbare earl succeeded in evading the ignominy of stooping so low as to ask for money for himself by claiming that the proceeds from his hospitality were going to "charitable purposes." Visitors were admitted to the grounds on payment of three shillings, an exorbitant charge considering how many thousands of people flocked to see the gardens.

It was not only the opening of the castle grounds which brought William Barron to the world's attention. In 1852, shortly after the death of the fourth earl, Barron wrote a successful book, *The British Winter Garden*, which met with

The pleasure grounds at Elvaston Castle, showing the clipped yews arranged in position by Barron's tree-moving machines, which made it one of the first ever "instant gardens." (*The Gardeners' Chronicle*, 1891)

ready sales. Full of valuable hints as to the proper method of moving trees and the right replanting conditions, it also dealt with difficulties resulting from planting "pot-bound" trees, which up to that time had not been much appreciated. He complemented his considerable knowledge with striking illustrations. Barron soon became one of the most prominent figures in the horticultural world, not only as a landscape gardener but also as a great authority on conifers. The "Barron transporter" certainly had some disadvantages—it required regiments of labor, and because it carried the trees vertically the selected route had to be free of overhead obstructions. Even so it was widely used, and its design was emulated by other gardeners. At Edinburgh, a relation of William McNab, James McNab, developed a strengthened tree-mover along the lines of Barron's, as did Edouard André, who used Barron-style machines to reshape the tree-lined avenues in Paris prior to the 1867 French Exposition.

Leaving Elvaston in 1862, Barron went into business as a nurseryman and landscape gardener, establishing premises at Borrowash, almost within sight of the garden he had done so much to make famous. William Barron & Son (he was joined by his eldest child who had returned from studying gardening abroad) accumulated a large and varied collection of hardy trees and shrubs, especially conifers. Yet Barron could never be induced to sell his choicest specimens—which he called his "Decoy Ducks." A shrewd businessman, he knew they lured in the customers. His real renown derived from his great experience in the moving of large trees; he would undertake with confidence tasks that others would—and did—shrink from. The results he achieved in terms of arrangement and high rates of success were without parallel. He took on numerous commissions all over Britain, so that he might have projects under his supervision in a dozen counties at once.

The transporting of an ancient yew in the churchyard at Buckland-in-Dover in 1880 became by far Barron's most famous achievement. Mentioned in the Domesday Book and reputed to be over a thousand years old, the much-loved tree had grown too close to the west wall of St. Andrew's parish church and had to be moved. Amid considerable local trepidation, in 1879 Barron was invited to give his opinion on the likelihood of successfully transplanting the tree. There were many who objected to its removal but, undaunted, he declared the move feasible. The undertaking excited worldwide interest, attracting the press and many writers. Hundreds of members of the public flocked to

William Barron, pictured standing astride the trench, after the successful completion in 1880 of the moving of the famous "Buckland yew," a tree mentioned in the Domesday Book. (*The Gardeners' Chronicle*, 1891)

witness the removal, a privilege for which Barron—perhaps mindful of the fifth earl's venture—insisted each be charged two shillings and sixpence! Dug around, then raised by powerful screw-jacks onto rollers, the 55-ton ancient tree was slowly hauled by giant windlasses—such as were used for moving bathing machines—203 feet across the churchyard, successfully reaching its new site to much acclaim on March 5, 1880. The scale of this operation was probably never matched and Barron, who had been rather more apprehensive than he let on, later admitted that all the other trees he had moved had been "chickens compared to the Buckland Yew."

Enjoying perfect health, Barron kept himself active in the tree-shifting business until his retirement in 1889. He died on

April 8, 1891. To the end, his wonderful memory enabled him to recount the minutest details of events of decades earlier, and he was able to both read and write without glasses to within a week of his death. An altruistic entrepreneur, he remained a religious man of the sternest integrity, who under no circumstances would allow his principles to be subservient to his interests. His obituary in *The Gardeners' Chronicle* described him as someone "who while he lived *by* horticulture, never forgot that it was his duty to live *for* it as well."

 Visit the gardens at Elvaston Castle Country Park, Borrowash Road, Derby DE72 3EP. Website: www.derbyshire.gov.uk.

Carl Nyberg's Flame-Gun

In the spring of 1955 garden weeds surviving in the crevices of paths had as much reason to cringe in terror as those World War I trench soldiers who caught a glimpse of the new mechanical weapon—the tank—thundering toward them. This time the newfangled attack weapon belching fire was a hooded flame-gun on wheels—an innovative development of the blowlamp invented in Sweden by Carl Nyberg.

It has sometimes been said that Max Sievert, a Swedish manufacturer and exporter, was the inventor of the flame-gun, but the credit should really go to his engineer, Carl Rickard Nyberg. Born in 1858, Nyberg had diligently struggled to find a means by which a portable stove-like contraption, fueled by kerosene, could be made to produce a far more searing flame than would be required for cooking. By 1881 he had invented a simple method which he successfully put into practice with a working prototype. Oil was fed under pressure to a hot coil, or plate, where it vaporized and ignited into a very hot flame. Unfortunately for Nyberg, his financial overseer Max Sievert claimed the manufacturing rights, and went on to patent the design in 1887—by which time it was already being produced in Stockholm. Initially the popular usage for that gas *blåslampan* was as a plumber's or painter's hand-held blowtorch.

All the while there were other rivals to contend with. One, Frans Wilhelm Lindqvist (1862–1931), claimed to have made the first recognizable working paraffin blowlamp. However, whereas Nyberg and Sievert concentrated on the artisan market, Lindqvist focused his attention on developing a pressure stove. Patented in the late 1880s, the new stove was dubbed the "Primus" and eventually over fifty million of them were sold around the world.

Nyberg's flame-gun took a surprisingly long while to be put to use in gardens. Partly this was because, being gas-fueled, such contraptions were not entirely safe to operate near the body. Earlier types were used in the same style as a knapsack sprayer, or were moved on rollers. Nevertheless, by the mid-1950s their evolution had been such that the apotheosis had been reached, represented by the long-handled hooded machines whose great virtue was that their shields concentrated the scorching hot flames. These new contraptions were fitted with wheels, which could be readily detached to make the machines easier to use around buildings, on sloping banks, or in other awkward positions.

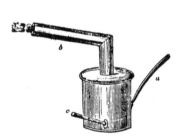

Until the development of the blowlamp in the 1890s, most gardeners used a fumigating pot, burning tobacco, in order to destroy grubs in greenhouses.
(Loudon, *The Suburban Horticulturist*)

Wheeled versions of the flame-gun variant of the blowlamp were on sale to gardeners in the mid-1950s. (*The Gardeners' Chronicle*, April 9, 1955)

Hooded machines could also be used safely between crop rows. They were able to destroy diseased material in orchards or in greenhouses, as well as burning old tulip and daffodil foliage. They were widely used for starting bonfires, and were invaluable for battling against unwanted soil dwellers such as slugs, wood lice, and flea beetles. It was not only *aficionados* of these Nyberg-derived machines who assumed they must be valuable for cleaning mushroom houses, poultry sheds, and pigsties. Buildings, cold-frames, and seedboxes were all thought to be made hygienic by such "flame brushings." At the time there was a fashionable belief that for plants to grow successfully it was essential for gardens to have "hygienic" soil and total freedom from weeds.

Although extremely fun to use, the flame-guns suffered from several disadvantages. The light "flame brushings" of the soil needed to be done with much care because if the flame was left in a stationary position for too long it would damage the humus or organic matter in the soil. The process was time-consuming and needed to be done when the weeds were small. It was unreasonable to expect the machines to immediately burn up large green weeds wet with dew or rain. The destruction of larger weeds or potato stalks was a two-stage job, involving killing first and then burning up a day or two later. Furthermore the "flame-brushing" left unsightly burns, and the noxious gas fumes did environmental damage to the air and even to the soil itself.

In the years after Nyberg's initial invention the main adaptation of the blowtorch from plumber's tool to weapon of destruction was carried out by Richard Fiedler, who in 1900 devised the flame-thrower. The resulting weapon, the *Flammenwerfer*, was used by the German military in devastating surprise attacks against the French lines in 1915. During World War II some tanks were equipped with flame-throwing guns.

Carl Nyberg himself died on March 25, 1939, just a few months before the outbreak of that conflict. The peace-loving Swede would no doubt have been saddened that his invention had not been confined to lightly burning weeds in gardens.

 The Swedish Inventors Association can be contacted via the Association of International Inventor Organizations, www.patentcafe.com.

PLANT FINDERS

🌸 Pietro Matthiolus's Tomato

This is an extraordinary story of how one of today's most commonly grown garden food plants was demonized in the medieval period, and of the influential physician who mis-classified it.

Wild forms of tomato probably originated in Peru, although the indigenous peoples of Mexico are believed to have been the first to domesticate them. Exactly when, or indeed how, the tomato reached Central America from the Andes is unknown, but by 1521, when the conquistador Hernán Cortés stormed Mexico City, the name *tomatl*, derived from the Nathuatl language of the Aztecs, was already established. The tomato's migration across southern Europe appears to have begun in 1523 when a Spanish explorer brought it back to Spain from Mexico, whereupon a Moor reputedly took it to Tangiers, from which point an Italian conveyed it to Italy. Alternatively it is possible that it went almost directly to Italy, because in the 1520s Naples was under Spanish rule. Perhaps the former explanation is the more likely because in Italy the new tomato was called the

Pietro Matthiolus. (Reproduced by permission of the Royal Botanic Gardens, Kew)

Moor's apple, *pomo dei moi*. It also acquired a name with a similar sound but a very different meaning: *poma amoris*, or "love apple." The French also referred to it as a love apple, *pomme d'amour*.

Pietro Matthiolus is significant because in 1544 he wrote the first printed reference to the tomato. Born in Sienna on March 12, 1505, he grew up to become a physician, like his father. Completing his studies in Rome, he worked as a medical doctor in Trento before moving on to become the city physician in Gorizia. His competence as a public health official—a job that would have brought him into close contact with the illnesses of the poorest people—was so highly regarded that he also became a medical consultant to quite a few members of the aristocracy. His renown was enhanced by the publication of several of his learned writings. The most famous of these was his edited version of a work by Dioscorides, *Commentarii Dioscorides*. Published in Venice in 1544, it listed all the plants known to Matthiolus, including some new discoveries he had made in the Tyrol together with descriptions of others sent to him by correspondents.

In the *Commentarii Dioscorides* Matthiolus described the tomato as follows: "Another species of mandrake has been brought to Italy in our time, flattened like the melerose and segmented, green at first and when ripe of a golden color, which is eaten in the same manner as an eggplant—fried in oil with salt and pepper." Although Matthiolus did not state that the tomato was poisonous—indeed he even described how it could be prepared for eating—his identification of it as a species of mandrake implied that it was. Many people, particularly in the Mediterranean region, regarded the mandrake with considerable superstition and fear, perceiving it to be both haunted and lethal.

Unfortunately for the tomato, the *Commentarii Dioscorides* was a phenomenal success. It became one of the most widely read books on botany of the period. Translated into several languages, it appeared in sixty editions (the first illustrated one in 1554), and made Matthiolus's opinions available to every student of botany and medicine in Europe. To make matters worse, in the sixteenth century European botanists assigned the pejorative name *Lycopersicon* ("wolf peach") to the tomato because—they claimed—it looked as inviting as a peach, yet was unfit for consumption. All this meant that for centuries the tomato was shunned. Insofar as it was grown in European gardens, it was really only used as an ornamental plant, perhaps hung in clusters from arbors.

The tomato's rehabilitation as an edible item began in America, where in the 1780s Thomas Jefferson planted them at Monticello. But the person most responsible for the sea-change in attitudes was Robert Gibbon Johnson. In 1820 this wealthy eccentric daringly gathered a crowd at Salem, then ate a basketful of fresh tomatoes from his own garden. He suffered no ill effects—although some of the flabbergasted spectators were reported to have fainted! This demonstration rapidly led to the widespread acceptance of the tomato as safe to eat, but commercial exploitation was delayed because of legal doubts as to whether tomatoes were fruits or vegetables; use in a stewing pot or on a salad plate suggested that the tomato was a vegetable, but botanically it was a fruit. In 1893 the U.S. Supreme Court ruled in favor of the tomato being legally classified as a vegetable, and therefore, to the delight of the U.S. growers, all imported tomatoes were taxed from that point on.

The *Commentarii Dioscorides* had brought Pietro Matthiolus such recognition that in 1555 he was summoned to Prague

and appointed physician to the household of Archduke
Ferdinand. It was ironic that Matthiolus, who was so concerned
with public health, had by his erroneous classification been
responsible for depriving the public of any chance of eating
nourishing tomatoes! He died in 1577—of plague.

 The American National Agricultural Library website contains
several interesting references to tomatoes: www.nalusda.gov/
pubs-dbs.

Mark Catesby's New World Depictions

This is the story of an unjustly forgotten hero. An
influential naturalist, pioneering environmentalist, and
author of the illustrated *Natural History of Carolina, Florida
and the Bahama Islands*, Mark Catesby facilitated the
introduction of many significant American trees and shrubs
to European gardens.

Mark Catesby was born in Sudbury, Suffolk, probably in
1679. From his earliest years he was adventurous in spirit
and possessed infinite determination plus a profound love of
nature. His affability and mild-mannered qualities encased a
robustly decent personal integrity. Much encouraged by
John Ray, the author of the celebrated *Historia Plantarum*,
who lived in the neighboring county of Essex, Catesby
applied his especially systematic mind to studying natural
history. As a student in London, he became fascinated by
stories of the developing colonies in North America, and in
April 1712 he leaped at an opportunity to explore those
new lands by sailing to visit his married sister and her
family in Williamsburg, Virginia. Many naturalists might

have regarded the voyage there as uneventful, but not Catesby. Always mentally alert, he missed nothing of interest and became intrigued by the movement of flying fishes.

Having an inherited private income meant that Catesby possessed the means to spend the next seven years as a hedonistic gentleman of leisure touring the American east coast, but instead he spent his leisure hours sending seeds of various types to friends such as the Hoxton nurseryman Thomas Fairchild. He carefully packed the samples in tubs and the shells of gourds, and also sent dried specimens. Catesby was exceptionally knowledgeable and methodical, although at that time he was just an amateur who gave no thought to doing such work professionally. He learned to paint, then traveled along the James River to its source, collecting plants, many of which he carried back with him when he returned to England in 1719. The seeds, field notes, and drawings that Catesby accumulated happened to be the largest natural history collection ever taken to England at that time, and it all meant he had unwittingly acquired a reputation as an authority on North America.

Back in England, some of the accounts Catesby had written about the natural history of Virginia had been read by a fellow of the Royal Society, the wealthy William Sherard, perhaps the most famous botanist of the age. He urged Catesby to return to America to gather material for a book on American natural history. Catesby was grateful for the acknowledgment and was certainly well disposed toward the scheme—but he was now financially less secure, his modest fortune having been dented by the South Sea Bubble fiasco. However, Dr. Sherard gathered the necessary funding from a syndicate of Royal Society members, including Sir Hans

Sloane—then Britain's greatest natural history collector—to whose museum Catesby presented a considerable number of specimens. Meanwhile, Catesby remained in England for a time, arranging and naming the specimens. In his written testimonial of the expedition Sherard described Catesby as a gentleman "pretty well skilled in natural history, who designs and paints in water colors to perfection." Then, the monetary support of his new-found patrons having been secured, in 1722 Catesby made a return trip across the Atlantic.

The altruistic nature of the sponsorship and Catesby's personal integrity became significant elements. Previously, insofar as there had been a few cursory reviews of the natural world of America, they had been for essentially mercantile purposes to chant a favorable litany of the New World's resources and products. In contrast, Catesby was under no such restrictions. Entrusted not to omit or embellish findings that failed to square with this perception, he was there to report accurately what he discovered and convey his honest impressions to those who supported him.

Catesby and the syndicate had decided that much of the fieldwork should be done in Carolina because its flora and fauna were relatively unknown (other than those elements related to commerce such as rice, oranges, and figs). Arriving in Carolina, which he lovingly wrote of as "a Country inferior to none in fertility, and abounding in a variety of the blessings of nature," Catesby made Charleston his headquarters. From there, for a year, he sallied forth exploring the inhabited section of the coast extending about 62 miles west from the sea. He went along the Savannah River, painting and collecting, and made several expeditions to the upper uninhabited parts of the country in the direction of the Appalachians. In spite of many difficulties, including the problems of obtaining suitable containers and packing materials, and his own ill-health, he traveled over a wide area. He subsequently wrote: "In these excursions I employed Indians to carry my box, in which besides paper and materials for painting, I put dry specimens of plants and seeds as I gathered them." Catesby was just as happy with Native Americans as he was in high society circles in London and, to his credit, he seems to have been one of the first European writers to give a favorable account of these people.

By this time Dr. Sherard and the other sponsors were becoming impatient for botanical specimens. To mollify them, Catesby sent a selection of dried plants and seeds which Sherard distributed among interested friends. One cargo destined for the good doctor was plundered by privateers and the vessel was forced to return to Carolina. One wonders what the privateers thought of their booty! Already Catesby was making his mark as a plant-finder. The native common catalpa tree, *Catalpa bignoniodes*, was totally unknown to the inhabitants of Carolina until he introduced it from remoter parts of the country, and by 1726 the plant was growing in England. Having made drawings and watercolor paintings of what he perceived to be most of the birds and reptiles of Carolina, Georgia, and Florida, in 1725 Catesby sailed for the Bahamas at the invitation of the Governor. He visited several of the islands, collecting plants, seeds, and marine products, all of which he dispatched to Sir Hans Sloane and other interested naturalists.

Completing his four-year peregrination in 1726, Catesby returned to England eager to settle down and convert his findings into the long-awaited book. Yet all was not well. In discussions with printers it became shockingly clear that Sherard had miscalculated the cost of getting the plates for

Catesby produced all his own illustrations for *The Natural History of Carolina, Florida and the Bahama Islands*. (National Trust Photo Library)

the *magnum opus* engraved. Even if such artwork were done in Amsterdam or Paris, the costs would be prohibitive. Sherard's death in 1728 came as a further blow (although the affluent naturalist Peter Collinson agreed to take over as principal sponsor). To make ends meet and get the project moving again, for a while Catesby worked for Thomas Fairchild's "City Gardens" nursery in Hoxton, which was famous for its American plants—partly due to Catesby's introductions! In due course he established his own commercial garden within the grounds of Fulham Nursery. Undeterred by the prohibitive costs of completing the illustrations, and with the same determination with which he had taught himself how to paint, Catesby took engraving lessons with the professional artist Joseph Goupy, and stoically set about etching the necessary two hundred copper plates himself.

The initial volume of the two-part *Natural History of Carolina, Florida and the Bahama Islands* was published in 1731, and the second, eventually, in 1743. With grand folio pages, sturdily bound in leather, it was the first large illustrated book on the fauna and flora of North America. Irrespective of the rather stiff, primitive quality of the execution, the pictures had a beautiful vitality, each one superbly colored by Catesby's own hand. The book was a testament to his integrity and thoroughness. It had been his custom in the field to sketch plants while they were fresh and just gathered. Even birds, and other wildlife that fed on such plants, he had invariably drawn while they were still alive. In the text (written in English and French) he pulled no punches when reporting the devastation he had witnessed already being inflicted on the environment. Ominously, he warned of the prodigious numbers of passenger pigeons being hunted to extinction on the mainland. Of the Bahamas he lamented that, although earlier in the eighteenth century the Bahamians had derived most of their principal income from the export of brasiletto wood, by 1725 those large trees had become uncommon in the islands, most of them already having been cut down.

The acclaim with which the *Natural History of Carolina* was received gained Catesby a worldwide circle of influential friends with whom he corresponded regularly, including the American naturalist John Bartram. In 1733 he was admitted as a fellow to the Royal Society. When his book was translated into German it brought him to the attention of continental readers.

Inexplicably, Catesby the dedicated scholar waited until he was sixty-four before he chose to get married—to a widow with a couple of grown-up children. In the meantime, down at Fulham Nursery, he developed his expertise in cultivating American plants. He noted that plants grown in moist soil in North America died when planted in similar soil in England; he deduced this to be due to the lack of sunshine in English

The title page of Mark Catesby's *Hortus Europae Americanus*.

summers. From his relatives in Virginia he constantly requested seeds of many kinds. All this data he used in his final book. *Hortus Europae Americanus* (1747) described eighty-five curious North American trees and shrubs which could adapt to the climate and soils of the British Isles and most parts of Europe. Catesby survived the completion of the book by some thirty months. When he died in London on Christmas Eve 1749 his plates and the few finished copies he had left were virtually his entire estate, and all that there was to support his family.

Many New World trees and shrubs were introduced into Britain and Europe by Catesby, including the catalpa. Significantly, as he observed in the *Hortus Europae Americanus*, the American colonies had "within less than half a century furnished England with a greater variety of trees than has been procured from all the other parts of the world for more than a thousand years." That book, and his *Natural History of Carolina*, added considerably to the introduction of such American plants into Europe, and they enjoyed an enviable popularity among amateur and professional naturalists throughout the eighteenth century.

Yet his reputation has subsequently languished. Although he was born in England and returned there to write his books, he did not seem to rate very highly as a British naturalist. Furthermore, although his work was almost entirely concerned with America, the consequence of spending just a few years there was that Americans regarded him as a transient visitor. A quiet, unglamorous, and upright character, Catesby was a diligent perseverer. He deserves to be remembered because he contributed so immensely to the Old World's knowledge of the New.

 A website showing an interesting selection of Catesby's pictures is www.philaprintshop.com/catesby.

John Bartram's Plant Boxes

Better known as America's greatest gardener, one of John Bartram's lesser-known accomplishments was as a pioneer of selling plants in boxes! His father and grandparents were Derbyshire Quakers who had journeyed to William Penn's new colony in 1632. Born there in 1699, John Bartram developed a prodigious interest in botany while growing up on his father's farm. After leaving school at the age of twelve, he taught himself classical languages, medicine, and surgery from books. Evidently he wished to become a doctor, but instead he embarked on a career in agriculture, after inheriting a small farm from his uncle. In 1788 he purchased some land at Kingsessing, along the Schuylkill River, near Philadelphia. A man of tremendous drive and energy, in addition to managing to make the ground there productive by draining its swampy parts, he built a farmhouse from stone that he quarried with his own hands. It was there that he lived with his wife Eliza Hunt (after her death he married for a second time, and eventually had eleven children). Bartram was an outstanding and talented farmer who produced exceptionally high yields from his fields of wheat, flax, oats, and Indian corn. He was among the first to practice farming techniques which are *de rigueur* today, such as rotating crops, letting land lie fallow, and using compost with manure and gypsum.

A cultured yeoman farmer, tough yet refined, Bartram had an altruistic outlook as well as an understanding of the healing power of plants which meant he was willing to function as a *de facto*, though unqualified, medical practitioner. He knew enough about the science of medicine to be able to prescribe herbal remedies for his ailing neighbors who were out of reach of any recognized doctor. According to legend the catalyst for his change of profession from agriculture to botany occurred while he was plowing a field. He saw a daisy and decided to take a closer look, exclaiming, "It's a shame that thee should destroy so many without being acquainted with their structure or uses." For all his sense of public duty Bartram was also a shrewd entrepreneur who realized that he could make higher profits from trading in plants than he could from farming.

With astonishing speed he laid out what was effectively the first botanical garden in North America. In 1730 six of his riverside acres were stocked with an unrivaled collection of New World flora. In terms of trade he would seem to have been supremely well placed, with his business situated on the edge of a huge unexplored garden covering thousands of acres, all of it largely unknown to Europeans.

The wealthy London wholesale haberdasher Peter Collinson provided Bartram with a unique means of forging revolutionary links with the gardens of England. Throughout Britain growing numbers of landowners were fascinated by the new varieties and Collinson guessed Bartram would be well placed to satisfy the demand. A keen naturalist whose hobby involved the search for new and exotic plants, Collinson's trade with America had brought him into contact with numerous settlers. However, none of his American customers was interested in sending him plants and seeds. Then, via Dr. Sam Chew, a Quaker friend of Benjamin Franklin's, Collinson wrote to Bartram asking him if he could provide plants. Thus began an extraordinary correspondence between the two men. They would exchange thousands of letters over the next thirty-seven years, until Collinson's death, but in all that time they never met.

Bartram quickly realized that here was a means by which he might receive a guaranteed income. Before this his American customers had often been slow to pay him for the plants he had supplied, and some never paid up at all. With Collinson he devised a radical subscription service, which he hoped might catch on in North America, whereby he would provide "garden-boxes" filled with 105 choice American seeds, roots, and plants.

These were shipped across the Atlantic to Collinson's houses at Peckham and Mill Hill from where they would be forwarded to customers. Bartram duly sent Collinson countless boxes, some of which also included curios such as tortoises, snakes, birds' nests, frogs, fossils, and even beetles. In return, over many years Collinson sent little money but he did dispatch goods in kind— clothes, books, and other supplies—and sometimes European seeds and plants to be tested out in America. By this means, with the assistance of Collinson, Bartram was reputedly responsible for introducing to North America vines, nectarines, almonds, the horse-chestnut, and the cedar of Lebanon. Their letters were full of accounts of safe arrivals, or of disasters— ships were delayed or plants dried out, and once rats nested in the boxes, destroying the precious contents.

Collinson arranged for the garden-boxes to be sold on subscription for a fixed annual fee of five guineas each. The subscription list is said to have grown from about fifty-seven to just over one hundred and contained many significant names including the Prince of Wales, the Dukes of Norfolk, Argyll, Richmond, Marlborough, and Bedford, and Philip Miller, Superintendent of the Physic Garden at Chelsea and author of the famous *Gardener's Dictionary*. One patron, Lord Petre, planted some ten thousand American trees. It was through Collinson's contact with such noblemen that in 1765 George III appointed Bartram the royal botanist for Florida for an annual stipend of £50.

With his extraordinary energy and determination, Bartram was an ideal choice for that position, especially as the plant-collecting elements of his job accorded so well with his particular attributes. He had courage, a strong physique, and was a shrewd and careful operator. He knew how to search for plants and in 1751 had published a book describing his methods: *Observations on the Inhabitants, Climate and Soil Made by John Bartram in His Travels from Pennsylvania to Lake Ontario.* He was observant and had a highly retentive memory, and it was claimed of him that not a thousandth of what he knew was ever written down. Always in search of new specimens for his garden, his frequent collecting expeditions meant he explored everywhere from Lake Ontario to Florida. He never forgot his passion for medicine, and that influenced his collecting zeal. New plants might mean new drugs. His peregrinations produced an astonishing assortment of new plants. Indeed, of the three hundred or so new plants introduced into England between 1735 and 1780, Bartram and Collinson were responsible for perhaps two-thirds of them.

Bartram was advanced in his views, both political and religious, and his outspoken opinions caused him to be disowned by his co-religionists in the Society of Friends. He demonstrated his opposition to slavery by freeing his slaves, which was a forthright stance in an age when abolition was unpopular. But Bartram was defiant. Having freed his slaves he worked alongside them in the gardens, and always had them at his table, both with his family and when entertaining his most distinguished guests.

Thomas Jefferson often bought plants for Monticello at Bartram's nursery, and George Washington was known to visit too. A naturalist as well as a botanist, Bartram also described and collected zoological specimens, proposed geological surveys of North American mineral sites, and argued that fossils should be investigated scientifically, rather than exploited as curiosities. With his growing international fame came accolades: Linnaeus described him as "the greatest natural field botanist in the world," and named a genus of mosses *Bartramia* in his honor.

Bartram's death was ironic in view of his devotion to the concept of plant exchanges between North America and Britain. In September 1777, just a week after British troops under the command of General Sir William Howe had routed George Washington's forces at Brandywine Creek, they approached Bartram's nursery. Fearing for the safety of his life's work, Bartram died on September 22, apparently of acute anxiety. Tragically, he was unaware that in view of his status as the royal botanist the British had been issued with an edict by the king not to disturb his farm and garden.

Visit the pre-revolutionary house of John Bartram: The Historic Bartram's Garden, 54th Street & Lindbergh Boulevard, Philadelphia; www.bartramsgarden.org.

Nathaniel Wallich's Rhododendron

Now so widely grown in England, the distinctively blood-red *Rhododendron arboreum* could perhaps be thought of as the most quintessential of English tree rhododendrons—but not so! Surprisingly, the person most responsible for introducing it into Britain—by packing it in tins of sugar—was a Dane. In addition to distributing many lesser-known Himalayan plants throughout the world via his famous Calcutta Garden, Nathaniel Wallich was also highly influential in providing a wider understanding of plants in India.

The historical significance of *R. arboreum* was that it was the first Himalayan species to be introduced to Europe. At the beginning of the nineteenth century fewer than a dozen exotic species were in cultivation, with mixed success, in the western world. Of those, the Alpine rose (*R. hirsutum*), introduced to Britain from the European Alps in 1656, was fairly insubstantial, and the common rhododendron (*R. ponticum*), discovered near Gibraltar by the Swedish naturalist Claes Alstroemer in 1750, proved to be a major weed in the forested parts of Europe because of its uncontrollable growth. As yet the most beautiful garden-friendly rhododendrons were still living wild in the Himalayas awaiting discovery.

Nathaniel Wallich was born in Copenhagen on January 28, 1786, the son of a Jewish merchant, Wulff Lazarus Wallich. Leaving school, Nathaniel entered the university at the early, but not then unusual, age of fifteen as a student of veterinary medicine and botany. Utterly fascinated by natural history, he had the good fortune to be taught by a Linnean disciple, Martin Vahl. It seems that his parents did not consent willingly to Nathaniel's choice of study, as they would certainly have preferred a more practical career for their son. After a short while he seems to have yielded to persuasion, because he reluctantly changed to medicine, graduating from the Royal Academy of Surgeons in 1806 as a Master of Surgery.

If his family expected him to settle down and start a practice in his native city they were surely disappointed. The young surgeon still showed strong—and adventurous—inclinations toward natural history. Having entered the Danish medical service at the age of only twenty-one he almost immediately got himself appointed as surgeon to the Danish settlement at Serampore, in Bengal, which he reached in November 1807. There he might have remained in obscurity but for a strange quirk of history.

During most of the Napoleonic Wars, Serampore (situated just north of Calcutta) had preserved a prosperous neutrality, but the bombardment of Copenhagen by the British in 1807

Nathaniel Wallich (1786–1854). (Reproduced by permission of the Linnean Society of London)

brought the fragile peace to an end, and in 1808 the colony was captured by British forces. Wallich became a prisoner of war, but his enthusiasm for and extensive knowledge of plants soon drew him to the attention of the British government in India, especially because at that time very few of the East India Company's servants had any knowledge of the subject. In 1809, through the intervention of colleagues, Wallich was released to assist William Roxburgh, Superintendent of the Calcutta Botanic Garden. Ironically, although he was a doctor, Wallich was himself of a somewhat frail disposition, and after an illness in 1812 he was required to undertake a sea voyage to Mauritius to regain his health. On his return he joined the service of the East India Company as an assistant surgeon. Wallich was the natural choice to succeed Roxburgh, who died in 1815, yet the company was initially reluctant to appoint a Dane to such a strategic position. Only in 1817 did Wallich—on the recommendation of Sir Joseph Banks and other prominent individuals—become permanent Superintendent of the garden, a post he held for the next thirty years.

There was some assumption that Wallich might be a rather learned Superintendent who would quietly edit William Roxburgh's posthumous *Flora Indica* for publication. However, his zeal as a collector of new plants was greater than his patience with working up existing material, and he was particularly skillful in preparing and transporting plants. He was soon sending large quantities of live plants, bulbs, and seeds to the botanic gardens at Kew and Edinburgh, as well as to influential figures such as Joseph Banks and William Roscoe. The spread of British political influence meant the army was opening up the countries bordering India, and the natural resources of those new lands needed to be accessed. Accordingly, in 1820 Wallich was officially directed to explore Nepal. There he undertook extensive travels, in the course of which he discovered and described many species of plants, some of them new to science.

In a just world, credit for introducing *R. arboreum* might have gone to an officer in the Indian Army, Captain Thomas Hardwicke. It is claimed that he discovered the tree species in its red-flowered form in Nepal in 1796 (although some people think it was 1799). He maintained that he found it flowering in Kumaon, southeast of Denta Dun in the Himalayas. An evergreen tree—Rhododendron means "rose tree"—of some 33 to 46 feet, it grew in a sheltered position. Beguilingly, *R. arboreum*'s colors varied with altitude. The deeper color forms were generally from lower levels and the paler ones from higher up. They were, however, temperamental, the rich color forms being noticeably less hardy than the paler ones and thriving only in milder climates. It is even possible that Hardwicke's original might have arrived in Britain in 1811—a curiously long time after being found. If so, it did not survive and needed to be reintroduced. The successful breakthrough eventually came through Wallich's excellent plant-handling skills. Realizing the importance of packaging, in 1821 he dispatched a consignment of precious *R. arboreum* plants from Nepal to the Liverpool Botanic Garden packed in tins of brown sugar! Though bizarre, this proved to be an effective method of protecting them through their long sea voyage. In 1825—a surprisingly short while afterward—came trustworthy reports of the first flowering of Wallich's crimson forms of *R. arboreum* in Hampshire.

Wallich put his zeal for collecting to good effect when he was sent by the British government in India to inspect first the timber forests of western Hindustan in 1825, and then the newly acquired Burmese territory in 1826. In 1828 the poor state of his health, which had become greatly impaired, required him to return to Europe. He took with him visible proofs of his never-tiring zeal in pursuit of science. Some eight thousand species of living plants and dried specimens, all collected by Wallich, safely arrived in London. The East India Company having been persuaded to grant consent, the plants were speedily dispersed to the public and private herbaria of Europe and America. Wallich spent the next four years in London, working on his *Plantae Asiaticae Rariores* (1830–2), a vast three-volume catalog of his studies of the flora of India and neighboring countries, illustrated with lithographs from drawings by native artists.

Wallich also had a lighter side. The center of operations in India, the Calcutta Botanic Garden, had originally been established in 1787 as a result of efforts by an army engineer, Colonel Robert Kyd, who had been so appalled at the widespread starvation caused in Bengal by a series of crop failures that he persuaded the East India Company to develop a suitable area of land in which to try out alternative crops. The company duly acquired 300 acres on the banks of the River Hooghly, an area subsequently used by Wallich as a receiving station, to which he encouraged local collectors to bring plants from regions not open to Europeans. In Wallich's time, in addition to their function as the largest botanical garden in Asia, the grounds achieved much renown as a pleasure garden and became known as "Wallich's Pet." A big-hearted man who was much in favor of public usage of scientific facilities, he welcomed its use by the Calcuttians for leisure purposes. The gardens were then in their heyday—indeed their golden age rapidly faded with the end of Wallich's term of office.

After a final expedition to make an extensive survey of previously unexplored Assam, with reference to the discovery of the wild tea shrub, ill-health overwhelmed Wallich. Leaving India in 1843 to seek a milder climate, he went this time to South Africa. This trip, undertaken for reasons of health, turned into a veritable businessman's holiday as he devoted all his energies to plant-hunting. In 1847 he returned to England with his family. Despite the skepticism with which the British authorities had regarded his national allegiances many years earlier, Wallich had long since come to view Britain as his homeland.

Wallich was not a man to rest on his laurels during the remaining few years of his life. The arranging of his immense collection and describing of new species kept him occupied all the time, and his endeavors in the field of botany did not go unrecognized. Though now in ill-health the great Dane received numerous prestigious honors, notably fellowship of the Royal Society and the presidency of the Linnean Society. Few garden pioneers enjoyed in their lifetime a higher reputation and esteem. He died in his sixty-ninth year on April 28, 1854, at his home in London, a man of warm affections and ready wit.

The Wallich name also lived on in other respects. His only son George Charles Wallich (1815–99), a surgeon with the Indian Medical Service, became an acknowledged pioneering zoologist. In the 1850s he was responsible for surveying the marine life of the North Atlantic floor prior to the installation of the transatlantic telegraph cable. In India Thomas Hardwicke overcame his disappointment at his failure to win the accolade of successfully introducing the first *R. arboreum*. He went on to have six children with his Indian wife, and became a major-general in the Bengal Artillery. It was

Sir Joseph Hooker who, following Wallich's footsteps in the eastern Himalayas during the 1850s, popularized rhododendrons in the western world, and introduced dozens of new species. In respect of Nathaniel Wallich's broad achievements, Hooker immortalized the great botanist by naming a palm *Wallichia*.

 For more details on the Calcutta Botanic Garden, see www.indianmuseum-calcutta.org.

"Chinese" Wilson's Regal Lily

The regal lily, beguiling white within, yet wine-colored outside, has become one of America's most popular plants. It was introduced by the "Indiana Jones" of plant-hunting, Ernest Wilson, who collected and introduced into cultivation a greater number of plants than any other collector.

According to parish records members of the Wilson family had lived in the Gloucestershire town of Chipping Campden since 1627. It was there, on February 15, 1876, that Ernest Henry Wilson was born, the eldest son of Henry Wilson, a railwayman, and his wife Anne (née Curtis), a florist, who both shared a great love of plants. On completion of his schooling Ernest began working for the Hewitt Nurseries at Solihull. Spotted as a promising young gardener, in 1892 he accepted a post at the Birmingham Botanical Gardens, Edgbaston. Notwithstanding the long hours of work there, he studied at the same time at the Birmingham Technical School, winning the Queen's prize. This prestigious award helped him to obtain, at the age of twenty-one, a position at the Royal

Botanic Gardens, Kew, where his ability was soon evident both in the lecture-rooms and in practical activities in the gardens. In October 1898 he gained a fellowship at the Royal College of Science, South Kensington, with a view to training to become a botany lecturer.

The circumstance which prompted Wilson to go to China was Dr. Augustine Henry's mention of the existence of the beautiful handkerchief tree *Davida involucrata*. A British physician in the Imperial Chinese Customs Service at Ichang, Dr. Henry had for years occasionally sent the Surrey nurserymen James Veitch & Sons descriptions and specimens of the astonishing floral wealth of the Chinese province of Hupeh. At that time trees were felled in great numbers for charcoal and Dr. Henry was concerned that so many trees there were disappearing. He urged Veitch's to send a skilled collector to gather seeds of the rare handkerchief tree. In 1898, upon the recommendation of the Director of Kew, Sir William Thiselton-Dyer, who had been asked to suggest someone suitable, Wilson was selected because he possessed strength, courage, and perseverance—necessary attributes for the plant collector in untrodden wilds. On making the appointment, Harry Veitch robustly told Wilson: "Stick to the one thing you are after and don't spend time and money wandering about." Wilson became the first known European plant collector to approach the East from a westerly direction. Stopping off briefly on his journey at Boston, Massachusetts, he received some invaluable additional advice from Professor Charles Sprague Sargent, the Director of the Arnold Arboretum and an expert on procedures for collecting and preserving plant material.

By 1899 Wilson had arrived in Hong Kong and was making his way to Ichang. En route he nearly drowned when a

E. H. "Chinese" Wilson. (Reproduced by permission of the Royal Botanic Gardens, Kew)

careless skipper capsized the Yang-tse riverboat he was traveling in. These were dangerous times for westerners daring to venture into China, as the country was on the verge of the Boxer Rebellion; Wilson himself was briefly imprisoned by warlords who suspected him of being a spy. Wilson's strength of character enabled him to survive these ordeals, as did his calm, capable, and good-natured personality which won him goodwill in every hamlet he traveled through. The sketch map given to him by Dr. Henry indicated the approximate location of one handkerchief tree in a mountainous area about the size of New York State! When Wilson eventually struggled to the spot he was devastated to find that all there was left was a forlorn stump. The only known handkerchief tree had been chopped down for wood to build a nearby peasant house! Such was Wilson's mettle that he did not languish in the depths of despair, or contemplate giving up the search. Heeding Veitch's words about sticking to one thing, he systematically searched the district, and a fortnight later he was thrilled to discover an entire grove of handkerchief trees in full bloom. That autumn he was able to gather the precious seeds he wanted. Luck had played its part. On subsequent journeys he saw other handkerchief trees, but none bearing seed. His objective achieved, he bought a houseboat to use as a base, then set about searching the mountains of north-western China for as many interesting new plants as he could find. Astonishingly he discovered nearly four hundred previously unseen types, specimens of which he safely brought home to Veitch's nursery at Coombe Hill in 1902, crammed into thirty-five cases.

That summer must have been a particularly happy time for Wilson. He was twenty-six years old and newly married to Ellen Ganderston of Edgbaston, who in no time it seems produced a daughter, Muriel Primrose. Later in the year Wilson was off again on his second Chinese journey on behalf of James Veitch. His task this time was to obtain seeds of the yellow poppy, *Meconopsis integrifolia*, which involved searching much farther to the west on the Sino-Tibetan frontier. Whenever possible Wilson traveled along the great rivers that cut through the mountains to get within trekking distance of the mountainsides and valleys which held the floral prizes he sought. Once again he succeeded. However, the poppy was probably one of the less exciting finds of the 1903 expedition as he also discovered many new rhododendrons, roses, and primulas. Of all the exotic finds of that expedition, Wilson never forgot his first glimpse of the white regal lily (*Lilium regale*) in a valley of the Min. He immediately realized it was a fine specialty plant with enormous market potential.

Meanwhile, the great firm of Veitch & Sons had fallen into a state of decline, a circumstance brought about by several tragic deaths in that family. Soon there would be no one able to direct the company. Furthermore, Wilson was crestfallen to realize that such had been the neglect in its horticultural care that quite a few of the seeds he had so valiantly searched for and conveyed from the East had failed to be germinated into seedlings, and some had even been dumped on the Coombe Hill rubbish heap. Because many of his tree and shrub introductions were neglected just as they were coming to maturity, they were never commercially grown or distributed. Much disillusioned, Wilson severed his links with Veitch & Sons and in the spring of 1905 took a post as botanical assistant at the Imperial Institute—just the sort of employment he had been about to undertake when his career as a collector began. It did not last for long. Within months he had been invited by Charles Sprague Sargent to make a third expedition to China, this time on behalf of Harvard University's prestigious Arnold Arboretum.

In 1907 Wilson started for China on a year-long trip with the express purpose of locating trees and shrubs suitable for the New England climate. The search covered the great mountainous region between China and Tibet—which was virtually in the same latitude as Florida. Wilson returned to Britain in the spring of 1909, but that September emigrated to America, now that he was a permanent member of the staff at the Arnold Arboretum. He did not take American citizenship, for he intended eventually to retire to Gloucestershire—although that was never to be.

While by nature he was not gratuitously finicky, Wilson was exceedingly determined and liked to have a task well completed. On his second expedition he assumed he had done enough by gathering specimens of the large and beautiful regal lily, but although Veitch managed to cultivate it in 1905, he had not developed it as a marketable product in Britain, and had completely failed to introduce it to America. Still rankled by Veitch's carelessness in not making better use of such finds, in 1910 Wilson set off on another Arnold expedition to obtain the lily in quantity. This time he approached China via Moscow and the Trans-Siberian railway. The valleys of the Min were steep and difficult to get to. Nevertheless the struggle to reach them was well worth the effort. Some were home to the most beautiful lilies, and curiously each valley had a species of its own. There, in the bare harsh conditions, surrounded by snow-covered mountains, Wilson found just such a valley teeming with the attractive trumpet-shaped white lilies. He waited until October, when conditions were ideal for digging, and gathered some seven thousand bulbs. They were soon packed and readied for transportation to America.

Since he was viewed as an important figure in China, ranking at least at mandarin level, Wilson was obliged to take a sedan chair with him as a mark of his status. It was a portable rattan structure which Wilson usually only ordered to be assembled for show when circumstances warranted. In effect it mattered more than a passport. Unpretentious and informal in manner, he actually preferred to walk with his thirty-strong team of Chinese attendants. That autumn, on the homeward descent through the Min gorges, where the cliffs consisted of mud shales that were liable to disintegrate after heavy rain, causing dangerous rock falls, Wilson, tired by the digging at high altitude, was using the sedan chair. Suddenly, without warning, the party was caught in an avalanche of boulders which came crashing down on the narrow path. Wilson struggled from the chair just before it overturned and slipped over a precipice, but a boulder caught him, shattering his right leg. A few feet away the servant carrying his wooden camera tripod was killed instantly. As the survivors tried to help the injured, bizarrely a mule train approached in the opposite direction. The track was so narrow that the hapless Wilson had to remain lying on the ground as, one by one, the fifty mules, their hooves as big as dinner plates, deftly stepped over him.

A splint was devised from the wrecked tripod and Wilson was carried for three days in the spare sedan chair on a forced march to the comparative civilization of a missionary station at Cheng-tu. His leg, lacerated and broken in two places, became infected and for six weeks the bones showed no signs of healing. Wilson later claimed he had been fortunate that the doctor there was a Quaker who tended the leg as best he could; anywhere else and the limb would have been amputated. After three months it had healed sufficiently for him to return, on crutches, to Boston where surgeons reset the bone. From then on, one leg was nearly an inch shorter than the other, giving him a permanent limp ("my lily limp," he called it). He had a special boot made which enabled him to cover many thousands of miles, but his limp prevented him from enlisting in World War I. In Boston and other American cities the episode captured the public's imagination and received much newspaper attention, earning him the nickname "Chinese" Wilson. He became a legend in the world of horticulture, and the coverage also provided the regal lily with tremendous publicity which helped to make it an international success.

Although delighted to be known as "Chinese" Wilson, Ernest was always diplomatically distant from the land in which he traveled. Invariably accompanied by a retinue of loyal interpreters and assistant collectors, he never bothered to learn the language. Humble, even when he wrote a string of much-acclaimed books about China—notably *A Naturalist in Western China* (1913) and *China Mother of Gardens* (1929)— he kept his readers' minds concentrated on the plants, and seldom provided any clue to his own adventures. Unlike bookish horticulturists who preferred to keep their subject to themselves, Wilson was outgoing and a delight to talk to. Those who knew him claim that even when showing total strangers around a garden he would be full of enthusiasm, instantly able to recall the Latin names of the plants, and good-naturedly willing to supply a stream of cheerful anecdotes of how and where such plants had been found.

Wilson never returned to China. Instead he began to travel farther afield, collecting plants in Taiwan, Korea, Australia, Tasmania, New Zealand, India, Singapore, Penang, Kenya, Zimbabwe, and South Africa. In Japan he discovered *Kurume* azaleas. When Sargent died in 1927, Wilson succeeded him as Director of the Arnold, where he was also given a doctorate. It was an extraordinary achievement for him to be appointed director at the age of only fifty, and a long and distinguished career seemed to stretch in front of him at the Arboretum, about which he had just written *America's Greatest Garden* (1925). Already he had introduced more than a thousand previously unknown species—many of which came to bear his name—and collected some sixteen thousand specimens with numerous duplicates. But everything came to a tragically shocking end on October 15, 1930, near Worcester in Massachusetts.

He was returning to the Arboretum with his wife after visiting their daughter Muriel in New York, when his car skidded on a road covered with wet fallen leaves; it smashed through a fence and plunged 49 feet down a steep embankment. Ellen was killed instantly and Wilson himself was fatally injured. He died on the way to the hospital without regaining consciousness. His reputation for physical indestructibility meant that the tragedy came as an almighty shock to horticulturists throughout the world. It was a tragic end for an intrepid explorer who had, for the sake of the regal lily and thousands of other plants, so nearly sacrificed his life in more hostile environments.

 Arnold Arboretum's website: www.arboretum.harvard.

WATER FEATURES

 Archimedes's Waterscrew

The inventor of the waterscrew is generally considered to have been Archimedes of Syracuse. Born in that important Greek city in Sicily in 287 B.C., he was the son of Pheidias, an astronomer. Archimedes studied with the followers of Euclid in Alexandria, where he won much renown as a brilliant physicist, and afterward in Sicily. Devoting himself to mathematical research he discovered formulas for the areas and volumes of spheres, cylinders, parabolas, and other shapes. He founded the science of hydrostatics, and gained some notoriety when running naked down a street shouting "Eureka!" having just formulated a radical displacement theory while in his bath. He also wrote on mechanical machines, notably the lever; he is reported to have said that with a long enough lever he could move the world. He set no value on the ingenious contrivances that made him famous, regarding such rewards as beneath his dignity as a scientist. In the service of King Hieron of Syracuse he devised a series of remarkable military engines to ward off a Roman siege of the city. These included long chutes which could be projected from the city walls for rocks to roll down, cranes with grappling hooks that could lift ships, and a "burning" mirror that could set the Roman ships on fire when they were within a bowshot of the wall.

By this time Archimedes had, it seems, devised his most famous invention, the waterscrew. The principle seems to have occurred to him several years earlier while he was a student in Egypt. On a barge trip along the Nile he is thought to have begun wondering if there was not a better means of raising water for irrigation than the conventional shadoof bucket pumps. He drew some diagrams and devised a wooden screw, like a long corkscrew, which was fitted into a cylinder-shaped case. At the bottom of the screw there was a handle. When the screw turned it raised the water from one spiral up to the next, by making the water move up the axis from crest

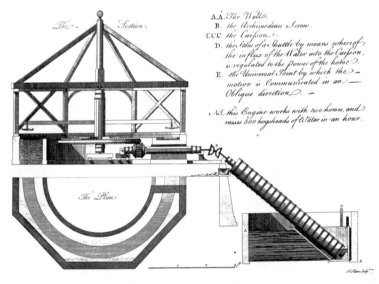

Two horses were used to power this sophisticated Archimedean waterscrew in the mid-nineteenth century at Kew Gardens. It could pump over two hundred thousand gallons in twelve hours. (Chambers, *Plans and Elevations at Kew*)

The principle of the Archimedean spiral was even applied to American mowers in the late nineteenth century and was advertised as being especially ideal for cutting lawns when wet. It was certainly easier to use than the patent turf cutter and combination garden implement! (*The Gardeners' Chronicle*, 1881)

to crest of the tread. Archimedes is variously claimed to have been asked by Hieron to apply the contraption to pumping water from the bilges of ships, and also to unloading grain from sea-going vessels in the harbor at Syracuse.

Variations of the waterscrew were produced. In addition to simply being a tube wound spirally about a cylindrical axis, it could also be a spiral revolving freely in a fixed cylinder. There is much controversy as to when waterscrews were first used in gardens. According to the *Hortulus*, written in 842 A.D. (and translated in 1966 by Walahfrid Strabo), machines resembling the Archimedean screw were used to lift water from the Euphrates to the Hanging Gardens of Babylon. However, it is not entirely clear when this took place.

As for Archimedes himself, during the siege of Syracuse he had exerted all his ingenuity in the defense of the city with remarkable success. When the Romans finally captured the city in 212 B.C. and massacred its inhabitants, Archimedes was struck down by Roman soldiers whom he rebuked while grappling with a mathematical theory. No blame for his death is attached to the Roman general Marcellus since he had given orders to spare the house and person of the sage. When Cicero, the statesman and eminent garden writer, was Quaestor in Sicily in 75 B.C. he discovered the tomb of Archimedes all overgrown with thorns and briers.

Subsequently waterscrews occasionally appeared in agricultural and horticultural settings. Before the eighteenth century, Archimedean screws were used in Holland and in the Fens of England for draining low-lying lands. Often they were powered by humans turning handles or treadmills, although sometimes windmills were used. The most sophisticated Archimedean garden of relatively modern times was built by John Smeaton at Kew Gardens in 1761. Powered by two horses under a 32-foot canopy, it supplied water to the lake and ornamental basins in Princess Augusta's garden and could raise some 3,600 hogsheads of water (approximately two hundred thousand gallons) in twelve hours. Smeaton is perhaps best known as the builder of the famous Eddystone lighthouse near Plymouth, but he also devised many innovative waterwheels. Yet even he, the foremost hydraulic engineer of his age, could not significantly improve on Archimedes's ingenious design.

 Useful website: www-history.mcs.st-andrews.ac.uk/history.

Thomas Hyll's Sprinkler

Although the first reciprocating pump using a piston in a cylinder was probably made in 250 B.C. at the time of Ptolemy by the ingenious Greek physicist Ctesibius, it was apparently not until the sixteenth century that the sprinkler was invented.

Thomas Hyll has the distinction of giving the first description of such a device, apparently by attaching a pump to a watering pot rose. In his book *The Gardeners' Labyrinth* (1577) he provides a section entitled "The Manner of Watering with a Pumpe in a Tubbe" in which he describes this watering process in detail.

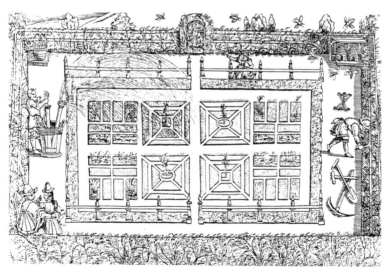

A sprinkler system for watering plants, as illustrated in Hyll's *The Gardeners' Labyrinth*. He seems to have been the first gardener to attach a sprinkling rose to a water pump for utilitarian purposes.

The vessel . . . must be set into a deepe vessell or tubbe of water, in what place of the garden the owner or Gardener mindeth to begin in drawing first the pump up, and with mightier strength thrusting it downe againe, which so handled, causeth the water to ascend and flee forth of the pipe holes on such height, that in the falling, the droppes come downe through the aire, breaking it in the forme of raine, that one place being sufficiently water, the Gardener may then remove the tub and vessel into another place, which needeth the like watering.

A maverick character who also used to write under the pseudonym Didymus Mountain, Hyll was probably born in 1543. A citizen of London, he was learned and well read in Latin, Greek, and probably also French. Having gained a doctorate, possibly at Oxford, he appears to have made his living as a hackneyed compiler of books, publishing numerous works on subjects as diverse as dreams, physiognomy, astronomy, and arithmetic, and also producing an almanac. Curiously, in some of his writings he describes himself as a man of "not much learning" and speaks of his "rudeness of pen." His last years were spent on the continent where he died in 1594.

Thomas Hyll (who sometimes called himself Didymus Mountain). (*The Gardeners' Chronicle*, 1874)

Even in Victorian times the sprinkler barrows—or "garden engines"—were still very much in action, and proved ideal for remote parts of the garden. (*The Gardeners' Chronicle*, June 1862)

Despite a somewhat buccaneering approach to his literary endeavors, "Doctor" Hyll certainly provided sound advice about watering plants. He is claimed to have had the distinction of writing the earliest English work on gardening, in 1563.

For further information on Thomas Hyll, see
www.hcs.ohio-state.edu/hort/history.

Ignace Dubus-Bonnel's Fiberglass Pond

In some small gardens now the largest constituent part—a fiberglass pond—is not only entirely artificially made, but its origins, surprisingly, are almost the most obscure. Fiberglass itself was invented a remarkably long time ago. A Frenchman,

Ignace Dubus-Bonnel, had deposited a patent as early as 1836 for the "weaving of glass, pure or mixed with silk, wool, cotton or linen, and made pliable by steam." By the means that he had devised, it was possible for imitation gold or silver brocades to be produced by combining glass fibers with metal particles. His fabrics won several prizes at the 1839 Paris Exhibition, and in 1840 the fiberglass inventor went on to produce the draperies that decorated the hearse used for the reburial of Napoleon's ashes.

The amazing fiber was virtually forgotten for the next ninety years, until 1932 when it was reinvented by an American bottle-maker, the Owens Illinois Glass Company. Initially it was manufactured on a large scale as a heat insulation product. Only in about 1938, when the Owens-Corning Fiberglass Corporation was formed, were rigid preformed products introduced. These were made by hardening the matted fiberglass with set resin. The improvements made during World War II in terms of technological advance and industrialization were such that increasingly fiberglass molds of all shapes and sizes were produced for and by the aircraft, boatbuilding, and construction industries.

This "wonder material," invented so many years earlier by Dubus-Bonnel, appears not to have reached garden suppliers until the early 1960s. Fiberglass ponds tended to be churned out at a rate of several a year by small firms making other fiberglass products such as parts for vans and cars. Each pool would be made by hand, taking about two hours to complete. After painting with resin, two layers of glass-fiber and polyester resin were rolled onto the mold. Once dry, the edges of the finished pool would be trimmed with a diamond cutter. Eventually a means was devised of using two molds, the liquid being injected between the two, thus avoiding the laborious hand-rolling.

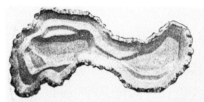

On the left is the *Trent*, a 12-foot, 130-gallon pool. Below is the 7-foot *Langdale*, holding 80 gallons.

Just two of the hand-built fibreglass pools designed by Minster with *your* profits in mind. See all the range of eleven, plain and rock finish, in our free 1970 catalogue. Write now to Europe's leading water gardening specialists.

MINSTER PROFIT-MAKERS 1970 STYLE!

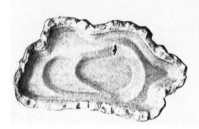

By the 1960s and 1970s fiberglass fishponds were being mass-produced, enabling owners of virtually any size garden to install an instant water feature in an afternoon—fiberglass itself had been invented as long ago as the early nineteenth century. (*The Gardeners' Chronicle*)

By then other alternatives were already available to gardeners seeking artificial ponds, such as non-reinforced plastic and pond-liner sheets made of butyl rubber or PVC. The advantage of the preformed fiberglass units was that they were light, strong, durable, and by far the fastest to install. They were available in a variety of shapes, sizes, and colors (garish shades of yellow, blue, and avocado were most popular) and all that was required was to dig a hole slightly larger than the pond, set it into place, level it, and replace the earth around the outside to fill in the gap. This could usually be done in one afternoon. Ignace Dubus-Bonnel could have had no idea that these molded versions of his invention would transform the appearance of millions of gardens.

For more about the making of fiberglass, see owenscorning.net.

🌺 Henry Bewley's Gutta Percha Hose

Surprisingly, what the development of garden watering equipment and the telecommunications revolution have in common is that they were both affected by Henry Bewley's invention in 1845 of a machine for making gutta percha rubber hose.

Insofar as there had been horticultural hoses for watering in ancient times, they appeared around 400 B.C. and, according to the Athenian painter Apollodorus, they were made of ox gut. What became the standard leather garden hose was invented by Jan van de Heide in Amsterdam in 1672. The early leather hoses were highly ineffective, being sewn together with boot-maker style stitches which easily broke when under pressure. Then in 1807 two members of the Philadelphia fire service, James Sellars and Abraham Pennock, devised a method of strengthening the leather by using metal rivets to bind the seams. Their new hose was such a success that they established a company to sell it throughout the United States. Nevertheless it had some serious disadvantages. Even when coiled, the standard 164-foot riveted lengths with metal couplings were immensely heavy, weighing over 79 pounds. The hoses needed constant attention and had to be regularly rubbed with neat's-foot oil to keep the leather soft and pliable. If it was allowed to become hard, cracks would appear and the hose would burst under pressure. Gardeners' apparel was often grease-stained from moving the oily hoses.

Gutta percha, the substance with which Henry Bewley would transform garden irrigation, had first been introduced to Europe by the Tradescants about the middle of the seventeenth century, although they did not really know what

Early hoses were made of leather and were produced in short sections. This ingenious barrow roller was typical of the designs of the day. (Loudon, *An Encyclopaedia of Gardening*, 1822)

Some hoses had rigid frames and needed to be trundled about on wheels. (Robinson, *Gleanings from French Gardens*, 1868)

it was. The younger Tradescant recorded it as being a unique form of "pliable Mazer wood" which softened in hot water. Then in 1843 an Indian Medical Service surgeon, William Montgomerie, who was serving in Malaya, noticed local people using gutta percha to make handles for their knives and tools. The substance was similar to india-rubber—indeed it was more pliable—but came from the sap of *Palaquium gutta* and *Palaquium oblongifolia* trees, native to the Malaya archipelago. It was obtained by cutting down the tree and then stripping off the bark, although the less destructive method of tapping could also be used. That year Dr. José d'Almeida delivered news of this wondrous substance at a lecture at the Royal Society of Arts, which subsequently awarded Montgomerie a gold medal.

Virtually no personal details are known about Henry Bewley, the chemist and entrepreneur who was so pivotal in the creation of gutta percha garden hoses. In September 1845, perceiving the substantial commercial potential in the creation of gutta percha products, Bewley obtained a patent for the manufacture of tubes, bottles, and hoses. He must have been an audacious risk-taker because at that time he had only just invented a workable contraption for making them! His extruding machine had a hopper down which the hot stock was fed to a piston and rotating screw which forced it forward through a nozzle of the required shape, the emerging length being cut off at desired points. The extruded lengths were coiled carefully in a bed of French chalk or other supporting powder to prevent its shape from distorting while it dried.

In 1845 Bewley formed his Gutta Percha Company at Silvertown in the London docklands and was soon selling the new type of hose. The tubing could be supplied in lengths of up to 328 feet, with union joints on the ends. Soon other manufacturers were producing elaborate water jets as sprinklers for lawn watering. The new hoses were certainly heavy, although by 1860 a wide range of inexpensive metal or wooden hose-holders on wheels were available in the form of trolleys, barrows, and carts, which were indispensable for moving the hoses without danger to plants or paths.

A patented extrusion machine such as this was used to manufacture the early gutta percha hoses. (Brant, *India Rubber, Gutta-Percha, and Balata*, 1900)

D'Almeida's lecture had been of particular interest to William Siemens, the electricity pioneer, who was seeking an effective insulator for submarine telegraph cables. At the Royal Society in 1848 the pioneering physicist Michael Faraday happened to suggest the use of gutta percha for this purpose, and in the same year a contract was granted to Henry Bewley for molding gutta percha and laying it in strips around a core. The first wire so insulated was laid across the Hudson River in 1849, and in 1850 the line from Dover to Calais was installed. Bewley's Gutta Percha Company

The gutta percha hose revolutionized garden watering and made it a pleasure everyone could enjoy. (By courtesy of the Plastics Historical Society)

became the Telegraph Construction and Maintenance Co. in 1864 and in the same year undertook to manufacture the first transatlantic submarine. Brunel's *Great Eastern*, having been chartered for the purpose, after several attempts successfully laid a permanent cable in 1866.

A disadvantage of the Bewley hoses was that in sunlight they tended to dry out and crack. This meant that gardeners had to treat them with oils—as they had needed to do with the leather pipes. The problem was partly overcome by mixing the gutta percha with india-rubber and rein-forcing the hoses with fabric. Nevertheless the introduction of artificial plastics even-tually led to gutta percha hoses being withdrawn. However, no better substance

J A M E S   L Y N E   H A N C O C K ' S
GARDEN WATERING HOSE and HOSE REEL,
VULCANISED INDIA-RUBBER HOSE PIPES, with
Union Joints for attaching to Cisterns or Water Mains.
Price per foot : —

Size.	½-inch	⅜-inch	¾ inch	Price of Brass Fittings.
Slight ..	0s. 6d.	0s. 7d.	0s. 8d.	Branch, 12-inch long, with Stop
Medium..	0 9	0 10	0 11	Cock, Rose, and Jet, 4s. each
Stout ..	0 11	1 1	1 2	Spring Spreader, 3s. each
For Mains	1 2	1 4	1 5	½-inch Union Joint, 3s. each

BRASS BENDS for Watering Plants in Greenhouses. These fit the usual Brass Branches.
For larger sizes and prices of Hose and Brass Fittings, see J. L. Hancock's Mechanical List.
The Hose Reel is a Simple Machine for readily Winding up and Wheeling away the Hose Pipes when out of use.
Vulcanised India-Rubber Works, Goswell Mews, and 266, Goswell Street, London, E C.

Gardening magazines advertised a wide range of accessories for hoses, including jet sprays and wheeled reel carriers. (*The Gardeners' Chronicle*, 1862)

was found for coating submarine cables. Indeed by the 1940s some 80 percent of the world's network of submarine cables had been coated with gutta percha produced by Bewley's extruding machine.

For more information on gutta percha, contact the Institute of Materials, www.materials.org.uk. The fire services were some of the earliest users of hoses. There is an impressive collection of old fire-fighting equipment at the London Fire Brigade Museum, 94a Southwark Bridge Road, London SE1 0EG; tel. 020–7587–2894.

GROWING EXOTICS

🌿 Salomon de Caus and His Orangery

The first recorded greenhouse was a movable wooden-framed structure erected at Heidelberg Castle in 1619 to shelter orange trees. Its builder was the ingenious Salomon de Caus—who nearly designed the first steam engine.

A native of Normandy, Salomon de Caus was born in 1576 in the town of Caux. He applied himself at an early age to the study of the mathematical sciences, his favorite writers being Archimedes, Euclid, and Vitruvius. After a visit to Italy he came to England as mathematical tutor to Henry, Prince of Wales. While in Richmond in 1611–13 de Caus built a gallery at Richmond Palace. He erected the south front of Wilton House, which was destroyed by fire in 1647, and designed elements of the gardens at Greenwich Palace and Somerset House. He seems also to have been employed as drawing-master to Princess Elizabeth. In 1613, after the death of the young Prince of Wales, de Caus was employed by the Elector Palatine, Frederick V, recently married to Princess Elizabeth, to rebuild the west side of Heidelberg Castle, creating the so-called "English Building," and also to

Heidelberg Castle, on whose grounds de Caus's revolutionary greenhouse was built. (Loudon, *An Encyclopaedia of Gardening*, 1822)

lay out the gardens there. This work would occupy de Caus for some years.

Since the sixteenth century, attempts by European aristocrats to keep up with the prestigious fashion for growing orange trees had frequently foundered owing to the inappropriate conditions in which the trees were kept in winter. Such arrangements involved the trees being brought in during the cold months and maintained in heated rooms which were very poorly lit. No wonder few survived. Realizing sooner than most that light was essential for the plants'

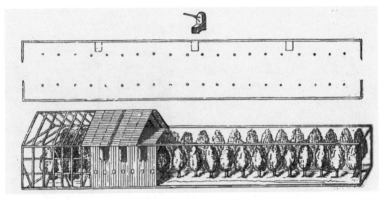

The world's first orangery with windows was designed by Salomon de Caus in 1619. A movable wooden structure, it was assembled each winter to protect the orange trees at Heidelberg Castle. (Loudon, *An Encyclopaedia of Gardening*, 1822)

nourishment, in 1619 Salomon built a giant barn-like shelter with glass windows to protect the Elector Palatine's four hundred orange trees—reputedly some of the oldest in Germany. It was 279 feet long, 32 feet wide, and high enough to take thirty of the largest trees—which were some 25 feet high—plus about four hundred others. A wooden-framed building, it was erected on site in the autumn and dismantled at Easter. It had four furnaces which were kept burning throughout the winter. Thus even during great frosts it was possible to walk in this orangery without feeling cold. At Easter the framework was dismantled, leaving the trees uncovered all through the summer. Because de Caus's structure was temporary, the orange trees could be left permanently in the ground and did not have to suffer the trauma of being moved.

In 1623 Salomon de Caus left the service of the Elector Palatine and returned to France, where in 1624 he published a work on sundials. His earlier book *Les Raisons des Forces Mouvantes avec Diverses Machines* (1615) became an important manual on hydraulics. Always devising whimsical machines for industry and human delight, de Caus also formulated a plan for a steam engine which, had it been put into practice, would have earned him the honor of inventing the world's first steam pump. Effectively he had figured out the scheme while engaged in creating waterworks and fountains such as those at Somerset House. There is an apocryphal story that de Caus lost his reason from the chagrin of being unable to convince Cardinal Richelieu of the importance of his discoveries, and while in confinement at the Bicetre hospital in Paris he was accidentally "discovered" by Edward, Marquis of Worcester, who extracted from him the secrets of his inventions and then published them as his own.

Salomon de Caus died in Paris on June 6, 1626. By then his temporary glass-paneled greenhouse had been such a success that permanent glass-windowed orangeries were already being built.

 Heidelberg Castle has been altered several times since Salomon de Caus built his greenhouse there but it remains a fascinating place to visit. Useful website: www.visit-heidelberg.com.

Henry Telende's Pineapple

Until 1720, when Henry Telende radically used tanner's bark to grow the first fruiting pineapple, these sensational exotics had been virtually untasted by Europeans. Such were the kudos and prestige associated with the pineapple that there were others—notably John Rose, the royal gardener—who sought to grab the credit for the achievement that was rightfully the unassuming Telende's.

The first encounter between a European and a pineapple occurred on November 4, 1493, when Christopher Columbus, on his second voyage to the Caribbean region, landed on an island in the Lesser Antilles, to which he gave the name Guadeloupe. Inspecting a deserted Carib village with his crew he found cooking pots filled with human body parts, and piles of fresh pineapples. Although the succulent fruits, with spiky dense heads surrounded by tufts of prickly leathery leaves, were a basic element of the diet of the inhabitants of tropical America, westerners had never seen the exotic fruits before and they were astounded by their fragrance and flavor. In 1563 the mariner Sir John Hawkins, while exploring the coast of what is now Venezuela, obtained pineapples from Carib Indians. He described the fruit as being the size of two fists and "more delicious than any sugared apple." Disregarding the Carib name "yayama," the Christians coined the name "pine-apple," since they thought it resembled a pine cone.

The Renaissance Europe to which Columbus had returned with his discoveries was a place where even the dullest fresh fruits were a rarity. The pineapple became an item of curiosity, partly because there was medicinal evidence that it prevented digestive disorders and was highly nutritious (on return voyages from the Americas sailors found that eating the pineapple could help prevent scurvy). However, such voyages were so slow, even if the fruit was picked when green at the point of departure, that by the time the Atlantic crossing had been made it was already rotting. This applied even more to pineapples imported from Singapore and southern India—where the Portuguese had introduced them. Insofar as "living" pineapples ever did reach Europe they did

This painting by Hendrik Danckerts supposedly shows John Rose (*c.* 1621–77), the king's gardener, presenting Charles II (1630–85) with the "First Pine-Apple Raised in England," *c.* 1670. (Private Collection/The Bridgeman Art Library)

so because they had somehow been brought over still attached to the entire plant, perhaps in a pot.

Despite the excited fascination with the pineapple, horticulturists had failed to grow the fruit in Europe, being unable to devise a means of creating tropical conditions hot and moist enough for them to flourish. Or so it seemed until 1661, when the artist Hendrik Danckerts was instructed to make a portrait of King Charles II posing in the garden at Downey Court with the horticulturist John Rose on one knee holding up a pineapple. The picture's title is *Rose, the Royal Gardener, Presenting the First Pine-Apple Raised in England.* The scene indicated how eager even the most senior public figures were to be associated with the pineapple's prestigious allure. Curiously the pineapple looks battered, almost as if it had been on a long journey. In fact the celebrated picture was a fake. According to John Evelyn's diary of 1661, the only event it was commemorating was the arrival of a potted pineapple from Barbados.

The gardener who actually raised the first pineapple in England—indeed, quite possibly in Europe—was Henry Telende. He thus became a significant garden innovator, although because of his humble origins very little is known of his life history. The gardener to Sir Matthew Decker MP of Richmond Green, Surrey, Telende happened to hear that tanner's bark had occasionally been used in hot-beds in Holland, though it was unknown in England, and had never been used for pineapples. The secret was that the oak bark tan used by leather-workers gave off tremendous heat when it fermented after being worked in tanning pits. Telende set to work with a 5-foot-deep brick-lined pit, 11 feet long and 7 feet wide. He put a layer of fermenting dung a foot deep at the bottom and covered this with 300 bushels of used tan bark, layering it with sprinklings of elm sawdust. The frame was covered with a glass light and two weeks later the necessary temperature was reached. Seeds and cuttings in pots were plunged into the bark, which maintained conditions at a steady, unchanging heat. In October that year Telende successfully ripened the first pineapple.

To record his achievement, and contradict John Rose's claim, Sir Matthew commissioned the artist Theodore Netscher to do a detailed oil painting of the fledgling pineapple. Its legend in Latin reads: "To the perpetual memory of Matthew Decker, baronet, and Theodore Netscher, gentleman. This pineapple, deemed worthy of the Royal table,

Henry Telende's pineapple, in a painting by Netscher, c. 1720. (The Fitzwilliam Museum, Cambridge)

grew at Richmond at the cost of the former, and still seems to grow at the art of the latter." Telende, the gardener whose triumph it really was, was totally ignored. It was only because in 1726 the writer Richard Bradley published an account of the methods used by Telende that the real hero was identified.

Telende's successful use of tanner's bark soon became widespread, not only in "pine pits" but also as a method of heating many greenhouses. It signaled the beginning of a period of intense interest in the growing of this plant on the large estates of wealthy landowners in England and on the continent. The gardeners on these various estates competed with one another in trying to produce superior fruits, and varieties not in the possession of other growers. An example of such was the Enville which was raised from seed at Enville Hall, the residence of the Earl of Stanford and Warrington, and grown in Florida and the West Indies. It produced a large, well-flavored fruit, characterized by a multiple crown—hence its local name, Cock's Comb.

A hostess's ability to produce a pineapple for an important dinner party spoke volumes about her rank and resourcefulness. So prized were the prickly fruits as centerpieces that some shops even rented them out to households by the day. By the late nineteenth century the pineapple's position as a status symbol had declined because swift imports from the Azores and canned varieties from Hawaii made it almost commonplace. That it had been such a status symbol, attainable in an edible form to just a few, was due to Henry Telende's tan-bark pit.

Netscher's original painting of the famous "Decker" pineapple can be viewed at The Fitzwilliam Museum, Trumpington Street, Cambridge CB2 IRB.

Chabannes's Greenhouse Central Heating

By the early nineteenth century the evolution of sophisticated glasshouse designs had still to be matched by the discovery of a practical means of heating them. That Holy Grail needed to be a safe, simple, reliable heating system which damaged neither plants nor buildings. Such a contraption came to be invented by the flamboyant Marquis de Chabannes, partly inspired by the invention of a French doctor called Bonnemain. Chabannes's hot water system was apparently so advanced that he claimed it could even be used in houses. It was, in effect, the first domestic central heating system.

Early in the previous century the ingenious horticultural equipment inventor Sir Hugh Platt had wondered if hot water might somehow be used instead of manure for heating hot beds and greenhouses, but he couldn't devise a way of doing it. The main disadvantages of conventional stoves, and especially open hearths, was that under glass their dry heat damaged plants by scorching and they gave off noxious fumes—and all that at an inconsistent temperature. Contraptions heated by steam (as opposed to hot water) became available from 1788, but these too had serious disadvantages: the vapor they emitted scalded plants, while their roaring boilers consumed copious heaps of coal and were a serious fire risk. Such boilers required a mechanic rather than a gardener to be on hand to tend them.

The concept of heating greenhouses by means of *circulating* hot water really originated from Dr. Bonnemain, who incubated chickens' eggs by manipulating hot water pipes into a radiator-like shape. During the fifteen years that preceded the French Revolution he busily hatched eggs on his small property outside Paris. It seems that he was one of the first to realize that the convective force of hot water was similar in some respects to the circulation of human blood. In 1777 he informed the Academy of Sciences about the "principle of heating by water circulation." Effectively the inventor of that notion, he went on to develop a boiler. Using various types of metal which expanded at different rates, he was able to automatically control the flow of air to the boiler, and thus the temperature it generated, to within half a degree. Dr. Bonnemain produced a pamphlet on the subject of hot water heating in 1816, but he never profited from his discovery. By the late 1820s, when he was more than eighty years old, he was reported to have lapsed into such an impoverished state that a request had to be made to the government to keep him from want.

Rather more is known about the history of the Marquis de Chabannes's ancestors than of his own early life. Since the twelfth century the family's powerbase had been the Château de Curton, a castle with extensive lands in the Dordogne area. Chabannes lost virtually all his fortune at the beginning of the Revolution when the already dilapidated property was confiscated and then sold off by the state. In 1803, apparently seeking to make a living as a property developer, the Marquis, now aged forty-one, published *Prospectus d'un Projet pour la Construction de Nouvelles Maisons*. He was known to be clever, and there was some doubt about his integrity as a businessman.

By 1810, and quite possibly earlier, Chabannes was living as an émigré in central London. That year he obtained a British patent for a hot water heating system with which to refine sugar. He claimed that the idea of using hot water rather than air in his *calorifère* suddenly struck him while he was fiddling with improvements to a hot water stove. It is just as likely that he was adopting Bonnemain's ideas. Another factor in his miraculously rapid metamorphosis into a central heating guru was that he was uniquely well placed to apply Bonnemain's ideas to Matthew Boulton's designs. An engineer who had been granted a patent in 1797 for raising water by impulse, in 1809 Boulton tried to devise a hot water

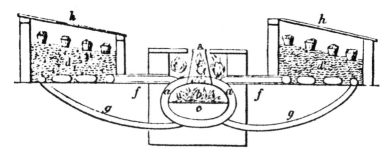

In 1818 Chabannes sought a patent for this fireplace and boiler designed to heat two or more hot beds. (Chabannes, 1818)

system for his own house in Soho. Within a year of Boulton's death later that year, Chabannes had somehow developed his own radical system enough to get it patented.

Chabannes's real break came at Sundridge Park, Kent, in 1812. In the grounds of the estate, which had been designed by the landscape gardener Humphrey Repton, Chabannes made history by fitting out a greenhouse with the world's first comprehensive water central heating system suitable for garden buildings. In the Nash-designed mansion there Chabannes also fitted a domestic central heating system. It was then that he made a disastrous mistake. Rather than developing his business's potentially highly profitable horticultural opportunities, the posh plumber decided instead to advertise central heating and air conditioning systems to owners of high-quality residential and public buildings. This shift was heralded in a sales pamphlet published in 1815 under his full, grandiose name, Jean-Frédéric de Marquis de Curton Chabannes: *Explanation of a New Method for Warming and Purifying the Air in Private Houses and Public Buildings: For Totally Destroying Smoke, for Purifying the Air in Stables, and Every Kind of Building in Which Animals Are Lodged.* From his eloquent writings at the time it is evident that his understanding of the subject was thorough, even to the medical elements. Indeed, some of his quack terminology was quite remarkably similar to that which Dr. Bonnemain might have used. In one place, for example, Chabannes claimed: "The fire is the power which gives motion to the water, as the admission of oxygen into our lungs causes the circulation of our blood. A pipe is placed at the top which may have any length or winding, but must finally return to the boiler."

Chabannes had a few successes. By 1817, having stressed the safety aspects of the system—it did not require a high-pressure boiler—he had installed heating systems in various high-profile establishments such as the Lloyd's assembly rooms at the Royal Exchange. At the Covent Garden Theatre, which had been burned down in 1800, Chabannes's system was chosen to warm the entrances and the famous stairway of the new building, thus helping to make the Royal Opera House the first centrally heated theater in Britain. However, it proved to be the last "garden" in which his system was applied.

By then he had established a manufacturing works at 121 Drury Lane, with an expensive showroom at 36–7 Burlington Arcade in London's fashionable West End. To stiffen the promotional effort, in 1818 he produced another publication, *On Conducting Air by Ventilation and Regulating the Temperature in Dwellings*, which was almost a work of genius. The booklet was illustrated with quarto plate diagrams of a garden hothouse, heated by a water-jacket boiler. Hot water was fed to radiant pipes running below the earth bed before returning to the boiler to be recirculated. However, the most fascinating—indeed futuristic—picture showed a complete domestic hot water system for a house. A kitchen fireplace boiler supplied hot water to a bath on the same level and to cast-iron heating urns on three floors above. To impress potential customers at his Burlington Arcade premises, at scheduled times of the week he would heat some rooms to exact temperatures, thus demonstrating that the system was controllable. Yet the buying public was skeptical of such stunts. They wondered if his showmanship might be a cover for some form of deceit.

Life cannot have been easy for the eccentric émigré attempting to sell a radical product far ahead of its time. So soon after the Napoleonic Wars the British public was still rather anti-French. Foolishly, Chabannes was too proud to compromise; high-handedly he persisted in calling his stoves *calorifères*. Even the 1818 pamphlet had echoes of a bitter aristocratic swansong. It concluded, "To have been in some degree useful to my fellow creatures is my ambition; to be one day accounted as such would be the reward of which I am most desirous." With just a hint of desperation he mentioned that "conservatories, hothouses and hot-beds have been heated upon this principle with the greatest success." But it was all too late. The business collapsed and Chabannes,

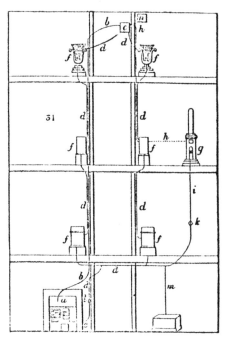

Chabannes's circulating system for a house, based on the circulation of blood in the human body. Arguably this was the world's first hot-water central heating system. (Chabannes, 1818)

ultimately ruined by his foolish strategy and extravagant operations, returned to France in 1821.

In the British provinces, where there was fierce competition to develop an entirely horticultural steam-free heating system, despite Chabannes's publicity stunts, astonishingly no inventor knew anything about the marquis's ready-made products. Even in London itself, when Sir John Soane was seeking a means of heating part of the Bank of England, he was quite unaware of Chabannes's hot water system. So in 1826 a paper was presented to the Horticultural Society which revealed that in 1822 one Andrew Bacon had devised the first hot water system on an estate in Aberaman, Glamorganshire. This was subsequently found to be void because the system did not have a return pipe, Bacon being unaware that warm water could circulate. However, the paper also brought to light the fact that another innovator, William Atkinson, had independently devised a circulatory hot water system (albeit later than Chabannes's). It was only in March 1828 that the real story of Chabannes's—and Bonnemain's—achievements was disclosed in an article by John Loudon for *The Gardener's Magazine*.

Crucially it was not until 1825 that Atkinson was able to make hot water circulate in pipes at a height of 39 feet. This he achieved by using a siphon mechanism, which had just been devised by Thomas Fowler. It seems that private customers had probably been wise in doubting Chabannes's exaggerated claims to be able to heat entire houses (although this did not invalidate the effectiveness of his apparatus on the ground floor of buildings, or in greenhouses).

Atkinson's system was less sophisticated than Chabannes's—which even had a water-jacket boiler—but it was he, not the French marquis, who won the historical credit for devising the first practical greenhouse hot water heating system. Yet Atkinson did not profit as much as he might have hoped, since numerous manufacturers were soon producing various applications of the hot water principle. A system developed by a company called Week's became the market leader. But that was no compensation to Chabannes who, virtually ignored, died in France in 1836.

 There are details on Chabannes's ancestral Château de Curton at www.castlesontheweb.com. For information on early domestic and greenhouse central heating systems, visit Sir John Soane's Museum, Lincoln's Inn Fields, London WC2; www.soane.org

George Stephenson's Cucumber

The brilliant engineer who built the first steam locomotive, George Stephenson was also a keen and innovative gardener who at his country estate, Tapton House near Chesterfield, devised a system for growing straight cucumbers.

Born on June 9, 1781, at Wylam, near Newcastle, the son of a colliery fireman, George Stephenson's first employment was herding cows. In character he was honest and sincere, and determined to succeed. From his earnings as an engine-minder he paid for an elementary education at night school. In 1814, while working as an engine-wright at Killingworth Colliery, he completed his first locomotive, which was used to haul a train to a port 9 miles away. Stephenson became a public figure in 1815 with his introduction of a miner's safety lamp—and the subsequent lengthy controversy with the supporters of Humphry Davy, whose own invention had appeared at the same time. In 1821 he was appointed engineer to the Stockton & Darlington Railway, and four years later the world's first passenger steam train in the world made its inaugural run. In 1826 Stephenson became chief engineer to the Liverpool & Manchester Railway. The triumph of his *Rocket* captured the imagination of the country and stimulated the great railway developments of the next decades, in which Stephenson took a leading part. In his prime he had exceptional strength and always took pleasure in simple and competitive outdoor recreations.

The cucumber originated in northern India and had been cultivated since earliest times. It spread westward to Europe and was raised by the ancient Greeks. The fruit was oblong, usually round, could be rough- or smooth-skinned, and varied in length from 4 to 24 inches. Low span-roofed houses heated with hot water pipes were generally used for their successful production in Britain. Loudon's *An Encyclopaedia of Gardening* (1822) listed fourteen principal varieties being widely cultivated. There was keen competition among gardeners to grow the perfect cucumber for exhibition at shows, but no matter what they did, the fruits were invariably distorted in shape.

During the great railway mania of 1844 Stephenson kept aloof from the mad schemes suggested by speculators, and used all his influence to keep the frenzy in check. According to his biographer Samuel Smiles, at Tapton House he lived the life of a country gentleman, enjoying his garden and grounds, and indulging his love of nature, which through all his busy life had never left him. It was not until 1845, following the death of his second wife, that he began to take an active interest in horticultural pursuits. Then he began to build new melon-houses, pineries, and vineries of great extent, and he now seemed as eager to outdo all other growers of exotic plants in his neighborhood as he had been some thirty years before to surpass the villagers of Killingworth in the production of cabbages and cauliflowers. Workmen were constantly employed enlarging his various greenhouses, until he had no fewer than ten glass forcing-houses.

Stephenson took no interest in flowers but he was intensely interested in fruit-growing, horticulture, and farming. He had special methods of feeding cattle and prided himself on his herd. He had his own ideas on the best shape for cattle, and their ribs (or "girders" as he called them), so that the maximum amount of weight could be carried. He also made experiments on the quick fattening of chickens. They were fed three times a day and after each meal were shut up in a darkened shelter to sleep. He made the radical observation that bees can carry more pollen if they are not required to fly uphill to their hives.

Stephenson was extremely successful in producing melons. He invented a method of suspending them in baskets of wire gauze, which relieved the tension on the stalk, allowing nutrition to proceed more freely. This enabled the fruit to grow and ripen more quickly. His grapes also took first prize at Rotherham, in a competition open

A gardening magazine advertisement for the type of cucumber glass perfected by Stephenson. (*The Gardeners' Chronicle*, 1857)

to the public. But he took most pride in his cucumbers. He raised them very fine and large, but initially at least he could not make them grow straight. Notwithstanding all his propping and humoring of them by modifying the application of heat and the admission of light, they would still insist on growing crooked. Then he had an idea. Several glass cylinders were made at his Newcastle steam engine factory, into which the growing cucumbers were inserted, and at last he succeeded in growing them perfectly straight. Carrying one of the new cucumbers into his house one evening, he showed it to a party of visitors; he told them how he had managed to grow it straight, adding "I think I have bothered them noo!"

He married for a third time in January 1848—on this occasion to his housekeeper—but then caught a fever and died at Tapton House on August 12, that same year.

For further information, see www.powerweb.freeserve.co.uk and www.lpl.arizona-edu.

Ephraim Wales Bull's Concord Grape

Until 1849 America had no great grape. There was no shortage of wild grapes in North America, but the settlers found their appearance strange and their flavor inferior to the established table and wine grapes of Europe. However, imported European grapes, born in milder climates, could not thrive in New England outside of a greenhouse and the grape never got off the ground as a New World crop.

The man who changed all that was Ephraim Wales Bull, the son of a Boston silversmith born in 1806. As a boy, Bull cultivated wild grapes in the backyard garden until he was apprenticed to a goldbeater at the age of fifteen. In 1836 he purchased seventeen acres of land near Concord, Massachusetts, and began growing grapes again. His goal was audacious: to grow a grape that could match the refined flavor of European varieties but was hardy enough to survive the bugs and killing frosts of New England.

Concord was an interesting place to live in those years. Bull's next-door neighbor was Bronson Alcott, the radical educator, social reformer, and father of Louisa May Alcott, who would in later years write *Little Women.* Just down the road lived Alcott's good friend, Ralph Waldo Emerson, the Unitarian pastor who inaugurated the Transcendentalist movement in 1836 with his landmark essay *Nature.* Henry David Thoreau, Emerson's most famous disciple, frequently visited the Emerson homestead in the early 1840s before moving into a cabin he had built himself on the shore of Walden Pond. When Bronson Alcott moved out of Concord in 1843 (to found a utopian vegetarian community called

Ephraim Wales Bull with his Concord grapevine. (Concord Museum, Concord, MA)

"Fruitlands") he was replaced by a struggling writer named Nathaniel Hawthorne.

As a uniquely American philosophy took root in Concord, stressing self-reliance and harmony with nature, Bull labored quietly in his vineyard to produce a uniquely American grape. To reproduce a grapevine by grafting is straight-forward; to breed a new vine is another matter. No two seeds are exactly alike, and the only way to judge the qualities of a seed is to grow it to fruition. Bull began by scrutinizing the wild grapes that grew on his farm. Behind his house he found one plant, a *Vitis labrusca*, that combined reasonably good flavor with the qualities he was seeking. He planted the grapes from this vine in the fall of 1843. On September 19, 1849, after cultivating the seedlings for six years, he picked a bunch of grapes that pleased him in their flavor and appearance. He planted the seeds from this vine, and the process began again. In 1853 he chose the best one and displayed it before the Massachusetts Horticultural Society. This hardy vine, which he dubbed the Concord grape, ripened early to beat the frosts and produced large purple-black fruit with a sweet, aromatic, full-bodied flavor. The grape, which was far superior to the native rivals such as the Isabella and Catawba, astonished the judges and won first prize. In his search for perfection Bull had grown some twenty-two thousand seedlings.

The Concord grape was a smash hit. In 1854 Bull began selling cuttings and earned $32,000 in his first year of business. There was every reason to think that Bull would be a very wealthy man, but it was not to be. Within a few years the grape had spread west to the Mississippi River and Bull's profits had slowed to a trickle. It turned out that other commercial nurseries had begun cultivating the grape and

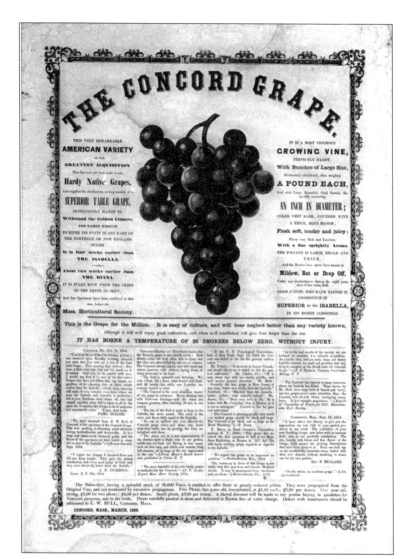

An 1859 advertisement of the Concord grapevine. (Concord Museum, Concord, MA)

selling it to consumers on the sly, cutting Bull entirely out of the loop. There was little Bull could have done to protect his creation. Plant patents were decades away, and it was impossible for Bull to sell the marvelous grape without also selling its seeds.

The man who ultimately would make a fortune off the Concord grape was a dentist named Thomas Bramwell Welch, a former minister and physician who created the first unfermented grape juice in his New Jersey home in 1869. A deeply religious man and a strict teetotaler, Dr. Welch was puzzling over a way to make a nonalcoholic communion wine. Welch harvested Concord grapes from the family trellis, cooked them briefly, strained the juice into bottles, sealed the bottles, and then boiled them to prevent fermentation (an application of Louis Pasteur's theory of pasteurization). The first batch was used successfully on the Communion table in the local Methodist church, and the processed fruit juice industry was born. At first Dr. Welch did not realize the potential of his new beverage, "Dr. Welch's Unfermented Wine," and cautioned his son to focus on dentistry, not juice. Early batches, which were very expensive, were used primarily for sacramental purposes. But prices came down and the juice caught on fast in a nation preoccupied with temperance. Welch's grape juice gained national attention when thousands tasted it at the Chicago World's Fair in 1893. The juice got another boost in 1913 when Secretary of State William Jennings Bryan (another teetotaler) made headlines by serving Welch's grape juice instead of wine at a high-profile diplomatic function. During World War I, Welch's introduced a new grape jam product, called "Grapelade," which the U.S. Army snapped up in its entirety and shipped overseas to feed the soldiers, many of whom returned home as devoted customers. In 1923 the company introduced its famous Concord Grape Jelly, although it was not until the 1940s that the company modestly suggested that the jelly might go nicely with peanut butter on white bread. Wines made with Concord grapes tend to exude a musky aroma and have never been popular, although the grape works well for sweetened wines such as Manischewitz. During Prohibition, the Concord grape was widely used to produce low-end bootleg wine.

Today, growers in the United States and Canada harvest more than 480,000 tons of Concord grapes each year. Welch's employs 1,300 people at its U.S. facilities and offers a wide selection of grape juices and juice mixes, jams and jellies, and fat-free frozen fruit juice bars, not to mention collectible jelly tumblers. In 2000 the company logged $679 million in sales. Not long ago—perhaps in a nod to its debt—Welch's relocated its headquarters to Concord, Massachusetts, not far from where Bull's farmhouse still stands on Lexington Road, and where the original vine that spawned the Concord grape still grows.

And what became of Ephraim Bull? He moved on with his life, although a certain sourness may have entered it. Soon after the Concord grape appeared he won election to the Massachusetts state legislature, and after that, held minor positions in Concord. He continued to grow grapes. In 1895 he died in poverty. The epitaph on Bull's headstone in Sleepy Hollow Cemetery offers the great man's last word on the famous grape: "He Sowed—Others Reaped."

For more on Ephraim Bull and the Concord grape, see www.concord.grape.org, www.national.grape.com, and www.welchinternational.com. For more information on the Concord Museum, see concordmuseum.org.

5

LAWNS

🌿 Pliny the Younger's Lawn

Although it is often assumed that lawns came into being in medieval times, their origins are actually far earlier, although whether the "flowery medes" apparently favored by the Persians and Mughals really constituted a grass lawn is most doubtful. What is certain is that the first person known to have owned a lawn in today's terms was Pliny the Younger.

From the year of his birth in 62 A.D., Pliny the Younger was brought up with an appreciation of beautiful gardens. Pliny the Elder was the author of *Historia Naturalis*, a vast encyclopedia in thirty-seven books covering natural history in all its phases. He even covered the subject of

Pliny the Younger was the first person known to have owned a garden with a lawn. (Mary Evans Picture Library)

The villa in Tuscany where Pliny had his lawn. (Loudon, *An Encyclopaedia of Gardening*, 1822)

how paths should be made. The commander of the Roman fleet at Pompeii, he lost his life when he went ashore on a rescue mission after the eruption of Mount Vesuvius in 79 A.D. Pliny the Younger inherited the family's estates near Rome and at the Tiberinum Villa in Tuscany, some 43 miles north of Rome, he established the first known lawn. The terraced gardens there covered about 24 acres and included a hippodrome, plus various theme gardens. The grassed area was a horseshoe-shaped sheltered walk which surrounded the garden's central features. In a letter written by Pliny to an associate, he described the lawn as a walkway enclosed with ornate evergreens shaped in a variety of forms. In another part of the garden was a small area of ground shaded by four plane trees, in the center of which was a fountain overflowing a marble basin. This was used to water both the trees and the grass.

In medieval Europe grass was used to provide a basis for sheltered bowling greens—remarkably similar to the covered

In Renaissance times lawns were often enclosed with flowery arbors, and many were used as bowling greens. (*The Gardeners' Labyrinth*)

walks at Tiberinum. Indeed, the word "lawn" did not appear in English until the mid-sixteenth century, and at first it meant just an open clearing in woodland. Thus, in both respects, Pliny might well have established a recreational lawn hundreds of years earlier than anyone else in Europe.

 See www.pompeii.virginia.edu; for further information on lawns and lawnmowers, visit the Milton Keynes Museum, Wolverton, Milton Keynes MK12 5EJ.

John Jaques II's Croquet

The cluck of mallets against wooden balls, the summer afternoon tea parties, and the smell of freshly mown lawns are the timeless images of the quintessentially English game of croquet. In fact the pastime was only established in Britain in the nineteenth century by John Jaques II, a Huguenot-descended Hatton Garden ivory turner, better known for manufacturing bowling-green equipment and indoor parlor games such as "ping pong" and "snakes and ladders."

Croquet's origins are obscure. The game is rumored to have been played in Ireland in the 1830s by French nuns using willow hoops and broomstick-handle mallets. It seems that John Jaques II happened to catch a glimpse of the game while he was traveling there on business. His father had established

John Jaques II, the innovator who popularized croquet. (Courtesy: Jaques & Son)

101

CROQUÊT:

THE

LAWS AND REGULATIONS OF THE GAME,

THOROUGHLY REVISED,

WITH A DESCRIPTION OF THE IMPLEMENTS,

ETC. ETC.

ILLUSTRATED WITH DIAGRAMS AND ENGRAVINGS.

By JOHN JAQUES.

LONDON:

JAQUES AND SON, 102, HATTON GARDEN.

1864.

[The Right of Translation is reserved]

He even wrote the rulebook!

John Jaques & Son, a high-quality games maker, in London in 1795. Sensing croquet's commercial potential, Jaques junior purchased the patent for the mallets and then introduced the game into England at the 1851 Great Exhibition, where his display won a Gold Medal. The widespread attention it attracted heralded the game's phenomenal success. Instead of retailing croquet equipment as individual items, the marketing genius John Jaques II sold complete kits in sturdy wooden boxes containing all the necessary accouterments—mallets, balls, hoops, corner flags, clips, and peg.

The game was played with as many as eight balls on a court containing six hoops and a center peg. Each ball had to run a set course, going through each hoop twice in a specific order and direction, before hitting the winning post. Enthusiasm for croquet was certainly not confined to the well-off country set who had the space for a full-sized (98 by 66 feet) croquet competition lawn. Indeed, provided that these proportions were roughly kept, it could be played on almost any lawn. This enabled the game to become popular with the aspiring middle classes who did not necessarily have large gardens. Shrewdly responsive to his customers' concern about the appearance of their lawns, Jaques wrote a glowing endorsement of some sage advice on the laying down of turf proffered by the famous garden guru Jane Loudon in *The Amateur Gardener's Calendar*.

> The best turf for gardeners is taken from fields or downs on which sheep have been pastured, as these animals destroy the coarse grasses by their habit of biting close to the ground. Frequent rolling in March and April is of the greatest importance; and if this is carefully attended to, and the grass regularly mown, a firm velvety lawn will be the result.

Another reason why croquet developed so rapidly was because it, like tennis parties, revolutionized Victorian outdoor social life by giving ladies the novel experience of playing a game in the open air in the company of gentlemen. It also provided an excuse for mingling and even for

A croquet ground. (Jaques, *Croquet*, 1864)

wandering off into the proverbial rhododendron bushes, momentarily out of sight of the ever-vigilant chaperones.

Despite its pejorative nickname "crinoline croquet," the game was invariably played with considerable exuberance and gusto. If a lawn was uneven, so much the better; it added to the knockabout nature of the game. Such was the lack of regulation that there were scarcely two lawns in England where croquet was played in the same manner in every respect. Lewis Carroll, an avid player at Oxford in the 1860s, captured the unruly nature of the game in his memorable passage of *Alice's Adventures in Wonderland* in which the croquet balls were hedgehogs, the mallets live ostriches.

Jaques's masterstroke was to compile an illustrated rule booklet, which from 1864 was packed into each box— thereby further enhancing his reputation as an expert on croquet matters. The guidelines were published annually and by 1867 had already run to 65,000 copies—an indication of the fortune Jaques was making from croquet. The instructions themselves were so liberal that they permitted, among other things, golf-type swings of the mallet, the putting of a foot on your ball before striking an opponent's, and hoops 10 inches wide—so wide, in fact, that a spaniel could run through them! Jaques also invented the technique of japanning the hoops white. This rendered them more visible against the green turf—a matter of some importance, especially when the game carried on in the dusk.

With its image as a game of Empire, by the 1860s croquet had reached Australia and New Zealand. It was played right across India, from Calcutta to Peshawar. A notable enthusiast was the Viceroy himself, Sir John Lawrence, who ordered Jaques & Son's top-quality set which had solid ivory mallets. The aura of Old Money and aristocratic nonchalance meant croquet was *de rigueur* in the elegantly fashionable parts of the USA's East Coast. In Boston, clerics and moralists even succeeded in getting the game banned on the grounds that it promoted licentiousness on the lawns—the sight of so many high society female ankles was deemed bad for the young men's blood pressure.

Ironically, the character most responsible for destroying the popularity of croquet was Walter Whitman Jones, an embittered gentleman inventor who had worked for Jaques & Son. Walter's younger brother Wolryche had devised a board game called "squails," then his elder brother Willie invented "the imperial Chinese game of frogs and toads." Both these board games enjoyed enormous success and were prominently advertised in John Jaques's croquet booklet. In contrast, "the game of war," a board game that Walter himself had devised, was a total flop. Furious with the manufacturers, Walter wrought his revenge by spitefully rewriting John Jaques II's croquet rules.

Astonishingly, in 1868 Walter's reforms were adopted by the All-England Croquet Club, a newly formed governing body based at Wimbledon. They totally transformed the game. Henceforth the emphasis was on complex rules, discipline, and shrewd tactical ability. Hoops were lowered in height to 12 inches, and drastically narrowed to 4 inches—just one-fifth of an inch wider than the standard wooden ball. On the other side of the Atlantic there was a furious reaction. Americans responded to an imperious attempt by the newly formed Croquet Association to ban innovative rubber-tipped mallets by engaging in a Boston Tea Party–style rebellion and establishing their own variation of the game, called "rogue."

Another consequence of converting croquet from a game of exuberance and spontaneity to a finicky pedantic sport—if not from being the silliest open-air game to the most intellectual—was that Walter Whitman Jones unwittingly made it vulnerable to competition from rival grass court games. Despite the fact that lawn tennis was patented by Captain Walter Wingfield in 1874 under the bizarre name of "sphairistrike," the new game was a phenomenal success. It could be played in a domestic garden as small as that which could accommodate a cramped croquet pitch, and had the advantage of being an energetic, fast, and free-flowing game. By 1877 the passion for lawn tennis had swept Britain, the Empire, and the United States. Many private lawns were converted to tennis courts and in 1875 even Wimbledon set aside some space for the new game. Although croquet remained a competitive sport, it was gradually replaced by tennis. In 1882 the Croquet Association, which Whitman Jones had been so instrumental in creating, was itself ignominiously expelled from Wimbledon.

For Jaques & Son the loss of enthusiasm for croquet meant a collapse in the sales of their boxed croquet sets. There was further disaster in 1941 when a bomb destroyed their Hatton Garden factory, and with it the records of John Jaques II's pioneering introduction of the game. However, the company was soon salvaged by British Intelligence who put it to work making secret escape devices to be included in Red Cross parcels sent to prisoners of war. Having survived croquet's remarkable ebb and flow of popularity, John Jaques & Son is now run by the fifth-generation Jaques.

 Contact the Croquet Association at The Hurlingham Club, Ranelagh Gardens, London SW6 5PR; www.croquet.org.uk. See also www.wimbledon.com and www.croquetamerica.com.

Charles Cathcart the Moss-Gatherer

In Victorian or Edwardian times any green sphagnum moss discovered growing on lawns would be promptly destroyed. However, since World War I sphagnum moss is much more likely to have been reverentially gathered, either to be chopped and mixed with soil in pots, or to be pressed into use as a moisture-retaining lining for hanging baskets. The forgotten hero responsible for this radical transformation in gardeners' attitudes was Charles Cathcart, the first doctor in the English-speaking world to gather moss to save lives.

Born on March 16, 1853, the son of a Leith wine merchant, Charles Walker Cathcart attended the famous old Loretto School where he excelled at sports, winning numerous medals in athletics and boxing. In the early 1870s, when reading for an arts degree at Edinburgh University, on three occasions he was chosen to play against England in Scotland's national rugby team. Belatedly recognizing that he had some talent as a physician he switched his studies to medicine. It was indicative of his quick learning powers that by 1878 he had been appointed resident surgeon at the Edinburgh Royal Infirmary.

Like several of the leading innovators Cathcart was deeply religious. Tall, self-effacing, and modest, he was extraordinarily energetic. His conscientiousness and devotion to duty were almost quixotic. A popular and excellent medical lecturer, he wrote *The*

Dr. Charles Cathcart. (*The Gambolier*, 1912)

Surgical Handbook, a paperback produced with specially rounded corners so that house surgeons could discreetly slip their copy in and out of the pockets of their white coats. Phenomenally successful, it was reprinted eighteen times. A man of great originality, he was always devising something new or producing a fresh idea. Endowed with considerable practical skill, he made papier-mâché casts that could be superimposed to give impressions of various layers of the body, and invented a simple apparatus to extend lower limbs during operations.

British surgeons had traditionally used absorbent cotton-wool swabs and bandages. But Cathcart, an expert clinical pathologist ever eager to push at the limits of existing techniques, endeavored to find more absorptive alternatives. As early as 1891 he began experimenting with other field dressings such as sponges, and even pads of pinewood sawdust. Then in 1900, while translating a paper on joint disease by a German professor, he happened to come across a remarkable story of the healing properties of sphagnum moss. In 1882 a woodman in a North German forest cut his forearm dangerously with an axe, and would have bled to death had he not staunched the wound with a handful of sphagnum moss. Some ten days later, when he arrived at a Kiel clinic with the original dressing undisturbed, it was feared that the wound would be septic when exposed to view. But, on the contrary, when the moss dressing was removed the wound was found to have healed almost completely.

Aware that folklore and other stories confirmed the extraordinary life-saving qualities of the moss, Cathcart enthusiastically conducted a few rudimentary experiments. He gathered 4½ pounds of fresh sphagnum, which he weighed out, then pressed. When the moss had been heat-dried he found it was able to retain nine times its own weight of fluid! Then, using a microscope, he discovered the secret of the sphagnum's power to absorb fluids—far more so than any other mosses. This was due to its capillary tubes, which were perforated cell structures readily able to take in water and hold it firmly.

Having been a commissioned medical officer in the Territorial movement since its formation in 1908, Cathcart was appointed chief surgeon of the leading Scottish Territorial Hospital in 1908 and was therefore already well placed to persuade the army to use moss bandages as cheap substitutes for cotton wool. At the outbreak of World War I Lieutenant-Colonel Cathcart hastily wrote a few letters to newspapers such as the *Scotsman* appealing to amateur gardeners to collect and send in the moss.

Initially packets and envelopes containing the moss arrived by ordinary letter post, but as the word spread the response from gardeners became overwhelming. In the summer of

1915 the collection of sphagnum moss became a mass movement. Because the moss was easily found and recognized, for women and children especially the gathering of it became a unique means by which they could assist wounded soldiers. In search of greater quantities of sphagnum, in the summer holidays moss-picking volunteers took to the Scottish Highlands. On September 14, 1915, the *Scotsman* reported a foraging scene at a West Highlands inn with freshly picked clumps of sphagnum being dried out on the lawn on sunny mornings: "The contrast between the scenes of peaceful beauty amidst which the moss is gathered and the war-ravaged sands for which they are destined is a most poignant one." Moss-pickers and gatherers were busy throughout the British Isles, notably in the Shetland Islands, the Lake District, and on Dartmoor. The central figure in all of this, Cathcart was called over to Dublin to organize the moss-gathering in Ireland.

Cathcart soon developed a simple industrial process by which volunteers could handle the moss. In various parts of the Scottish capital, in garden sheds, private houses, and church halls, working parties were established which would undertake the labor-intensive task of cleaning the moss, picking out any withered stems or other plants. It was then taken to the Royal Infirmary, where, because steam sterilization reduced the moss's absorptive qualities, it was treated with a mercuric chloride solution. Next the moss went to a factory where Cathcart had persuaded the owner to provide a hydraulic press; here the moss was squashed like cardboard into sheets measuring 12 by 20 inches before it was finally wrapped in muslin bags.

On the battlefield the advantages of the sphagnum moss bandages quickly became apparent. In its dry state it was

Edinburgh War Dressings Supply
(SPHAGNUM MOSS.)

Central Office—
37 PALMERSTON PLACE, EDINBURGH.

REPORT

AND

STATEMENT OF ACCOUNTS

From 4th November 1915 to 30th December 1916,

WITH

APPEAL FOR FUNDS.

COMMITTEE

The Hon. Lord MACKENZIE, *Chairman.*

Miss ASTLEY.
Miss BERRY.
Mrs BLACK.
Lt.-Col. F. M. CAIRD, F.R.C.S.E.
Miss CATHCART.
Lt.-Col. C. W. CATHCART, F.R.C.S.E.
Surgeon-General CULLING, C.B., D.D.M.S. Scottish Command.
Mrs CLYDE.

Sir JOHN COWAN, D.L.
Miss DALMAHOY.
Mrs LEONARD DICKSON.
Lady FAYRER.
Miss E. FINDLAY.
Rev. A. FORMAN.
Miss FRASER.
R. M'KENZIE JOHNSTON, F.R.C.S.E.

Miss LEECHMAN.
Miss LORIMER.
Miss M. R. MACLEOD.
VICTOR A. NOEL PATON, W.S.
Mrs ALISTAIR SINCLAIR.
A. FRANCIS STEUART.
Capt. W. J. STUART, F.R.C.S.E.
Mrs UNDERWOOD.
Miss E. WILKIE.

Joint Honorary Secretaries { W. A. PHILIP, A. W. SPALDING, } 37 PALMERSTON PLACE.
Joint Hon. Treasurers { A. G. MILLER, F.R.C.S.E. WALTER MACGREGOR, F.S.A.A., 35 ALVA STREET.
Hon. Auditor—WM. GREENHILL, C.A., 31 GEORGE STREET.

THE work of this organization, begun in April 1915 by Colonel Cathcart, R.A.M.C., and a few friends, on a somewhat modest scale, has now developed into an undertaking of considerable magnitude. As the utility of the moss dressings has become more and more recognised by Military Hospital authorities, so have the demands upon us increased. Fully 95 per cent. of our output is now sent direct to Military Hospitals at home and abroad, on the requisition of the D.G.V.O. at the War Office.

Apart from the great value of Sphagnum Moss as a surgical dressing, its highly absorbent qualities have proved of special service in another way. During the course of the present war a new method of applying antiseptics to wounds has been evolved

A report of the committee convened to organize production of sphagnum bandages.

111

extremely light, and so versatile and elastic that it could be formed into any shape. Its extraordinary ability to absorb fluids proved to be of special service in another capacity. During the course of the war doctors introduced a new method of applying antiseptics to wounds, known as the Carrel-Datkin treatment. Essentially, the treatment depended on keeping all the recesses of the wound constantly moistened with antiseptic lotion. When sphagnum moss was used to apply the lotion, the results were so astonishing that the method became the standard treatment for gunshot wounds in the British Army.

Large consignments of the bandages were distributed widely in Serbia, Gallipoli, Belgium, Salonika, France, Mesopotamia, and Italy, and they were hailed as a great success. Gathering together a group of public-spirited friends to form the Sphagnum Moss Committee, Cathcart launched an appeal for funds. The committee produced a report for the appeal in which it printed letters from battlefield surgeons in distant parts testifying to the success of the bandages. From Mesopotamia in 1917, one Red Cross commissioner wrote: "Sphagnum moss dressings have proved invaluable during the recent heavy fighting before the capture of Baghdad." With tragic irony, in June 1916 Charles Cathcart's only son had died of gunshot wounds in Mesopotamia. In his grief, the surgeon's response had been to throw himself into his war work. He honored the death of his son by creating a memorial garden in one of the poor sections of Edinburgh, but as the tragedy caught up with him he retreated from public life. When he died in 1932 his spirit was quite broken.

Voluntary labor during the war had made sphagnum bandages ten times cheaper to produce than the standard

So much sphagnum is now gathered from the wild for hanging baskets such as these that the environment is being damaged. (Author)

materials then in general use, but without that free labor supply they proved un-economical. So after the war the moss bandages ceased to be made. There was a resurgence of military interest in them in the 1940s, but they were quickly forgotten by the medical world. In contrast, the gardening world began to take an interest.

The increase in popularity of hanging baskets meant that many gardeners, unable to gather enough moss from their lawns, turned to other sources of supply. Many resorted to buying it in bags from garden centers, others to picking it in the wild. Because moss was not regarded as a protected plant it was often absent from the species lists of nature reserves. Conservation bodies gave out contradictory advice: some claimed that it was all right to pick it from roadsides, others insisted it was protected everywhere. That sort of confusion has led to its depletion as a wild species. For a while Colonel Cathcart gave people an affectionate understanding of the moss on their lawns, but today, sadly, it is only cherished for hanging baskets.

Sphagnum moss, and its botanical uses, is explained at www.botanical.com and www.ecn.purdue.edu.

Karl Dahlman's Flymo

The hovercraft was created by the brilliant pioneering inventor Christopher Cockerell. But unfortunately for the patriotic Cockerell, his invention was overshadowed by a phenomenally successful garden lawnmower variant of his idea—developed by a Swede. A prodigiously inventive Marconi electronics engineer who already had several patents to his name, Cockerell was living in a trailer by Oulton Broad in Norfolk in 1953 when he began to consider the possibility of travel by water-skimming craft. Others had dallied with the concept, but Cockerell took it a stage further. Using a pet food can inside a coffee can and an adapted vacuum cleaner, he developed a viable design in the form of a 24-inch model that could "hover" on a cushion of air.

Having solved the basic theoretical problem of developing a friction-free craft, Cockerell spent the next few years searching for financial backing; all the time he was fretting lest competitors in other countries—particularly Switzerland—might be developing a similar idea. By 1959 he had received just enough government funds to construct the SR-N1, a sea-going prototype. Needing to somehow attract British private investment in order to ensure future production he decided to engineer a few spectacular publicity stunts. In June 1959 the circular SR-N1 pugnaciously roared across the English Channel on its successful maiden flight with Cockerell eccentrically balancing himself on the deck as human ballast.

When just such a publicity photo—showing the nimble hovercraft in the Solent with the transatlantic liner *Queen Mary*—reached Sweden it inspired the imagination of

The SR-N1 hovercraft on its maiden cross-Channel trip in 1959. The eccentric figure on the foredeck, acting as movable ballast, is the hovercraft's inventor, Christopher Cockerell. (Hovercraft Museum Trust)

lawnmower technician Karl Dahlman. He knew that horizontally bladed rotary mowers, so popular among American gardeners, were heavy and difficult to maneuver. Might, he now wondered, a rotary lawnmower be devised which could hover on a cushion of air without the restrictions of wheels?

In several months of dedicated research and experimentation—albeit with an existing lightweight 50cc gas mower-engine mounted on top of a garbage can lid—Dahlman quickly found the practice of making a workable airborne lawnmower harder to master than the theory. Disappointingly the whirling horizontal cutting blades failed to provide enough air current to lift the entire apparatus.

Eventually he discovered the answer: attaching a second fan or impeller to the drive-shaft drew in air from above the hood, forced it out around the sides—and lifted the lawnmower. Unbeknownst to Dahlman, precisely this snag had been overcome by Cockerell several years earlier. Its design problems now resolved, the revolutionary new flying mower—looking astonishingly like a miniature version of the

Karl Dahlman, the inventor of the Flymo. (Electrolux)

SR-N1—was exhibited at the 1964 Brussels Inventors Fair. It won a gold medal and received international acclaim.

Contrary to Cockerell's expectations, Dahlman decided to manufacture the hover-mower in Britain, shrewdly realizing that this country had the largest market for lawnmowers in western Europe—and thus the greatest sales potential. His company, which he named "Flymo"—derived from the term "flying-mower"—began production in 1965.

The standard 15-inch Flymo was effectively the first-ever plastic lawnmower. Its lightness meant it was easy to carry and when stored in a garden shed it could be hung from brackets rather than taking up precious floor-space. With few moving parts it was reliable, and was guaranteed to last for years with only minimal maintenance. Lawns could be tackled whether the grass was wet or dry, and mowing over paths and paving stones presented no problem. The slimness of the hood allowed the mower to reach awkward areas, such as beneath overhanging bushes. Its versatility became especially apparent when cutting grass on banks whose steepness had previously made them inaccessible to conventional mowers.

Initially the revolutionary "flying lawnmower" was viewed by the gardening public with some skepticism but gradually Flymo won new orders by demonstrating its products on the doubting customers' own lawns. As ferry trips by hovercraft became commonplace, the more natural and readily acceptable became the idea of the hover-mower. Another factor was timeliness: the Flymo's unconventionality and ease of use accorded well with the liberal trend toward informal lawns, which needed to be tidy without being immaculate. Indeed, although the unevenness of the Flymo's cut and its inability to produce "stripes" threatened to make it vulnerable in the bitter Cylinder vs. Hover advertising war of the 1980s, it survived. Flymo's introduction of electric-powered and grass-collection models had closed the gap.

The prototype "garbage can lid" Flymo. (Electrolux)

Meanwhile, Cockerell, who might, had he wished, have been closely involved with the Flymo, preferred to plod on stoically with the development of the hovercraft. In spite of his efforts, the need for complicated airplane-style technology meant the costs of building and maintaining hovercraft were so uneconomically high that by the 1990s few hovercraft remained in use, and in October 2000 even the cross-Channel service was discontinued. Cockerell, whose hovercraft patents provided him with a living but never made him rich, went into modest retirement near Southampton—where he took up gardening. He died in 1999—the very year in which Dahlman's British-based Flymo group celebrated becoming Europe's largest manufacturer of lawnmowers and the UK's most popular lawnmower brand, in addition to selling well all around the world.

 Visit the British Lawnmower Museum, 106–14 Shakespeare Street, Southport, Lancashire PR8 5AJ, www.lawnmowerworld.co.uk, and The Hovercraft Museum, 15 St. Marks Road, Gosport, Hampshire PO12 2DA.

Frederick Law Olmsted's Great Lawn

Central Park was the first landscaped public park in the United States. It might also be described as one of the world's most ambitious and inventive gardens—an entirely artificial landscape of 843 acres, complete with forests and rolling meadows and brooks that babble at the turn of a spigot, walled in by some of the most valuable and densely populated real estate on earth. Of course, it wasn't always like that. In 1800 there were roughly sixty thousand people living in New York City, almost all of them in lower

Manhattan. By 1844 a wave of immigration had swelled the population to four hundred thousand, making New York the nation's largest city. That year the famed journalist William Cullen Bryant was the first to call for the creation of a large public park in Manhattan before the available land was used up. The largely upper-class supporters of the plan, looking with envy on the public grounds of London and Paris, argued that a comparable park in New York City would exert a positive influence on the workers and signal to the world that America had arrived. In 1853 city planners began acquiring an undeveloped rectangle of land between 59th and 106th Streets that contained little but swamps, rocky promontories, and a few quarries and slaughterhouses, although in the process they evicted some sixteen hundred people from shanties on the grounds of the future park, including a small African-American community called Seneca Village.

In 1858 the city opened a public competition to design the new park. The winning entry was a design entitled the "Greensward Plan" by Calvert Vaux, a British-born architect, and Park Superintendent Frederick Law Olmsted. Revered today as the "father" of landscape architecture in America, Olmsted at the time had little formal training that would have prepared him for such an undertaking. Although he had gained admission to Yale in 1837, sumac poisoning had weakened his eyes and prevented him from attending. Over the next twenty years, he tried his hand at a variety of professions—clerk, sailor, farmer, editor—but never completely succeeded at any. In 1850 he toured Europe and Britain, studying numerous parks and estates along the way; two years later he published his first book, *Walks and Talks of an American Farmer in England.* An opponent of slavery, he published three volumes of travel writing and social

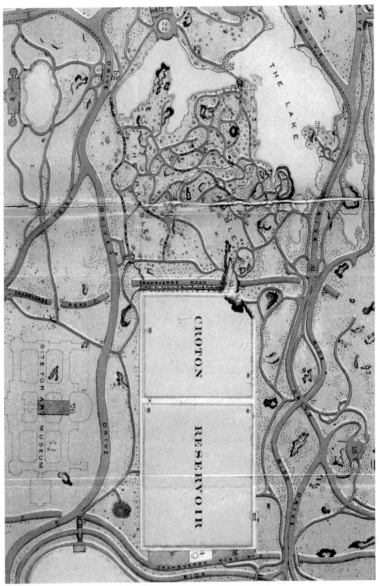

A portion of an early map of Central Park. (© The Central Park Conservancy)

commentary on the slaveholding South in the late 1850s. In 1857 Olmsted actively campaigned for the job of superintendent of Central Park and secured it the year before he and Vaux won the design competition.

Olmsted and Vaux's groundbreaking design combined open meadows and lakes with the picturesque intimacy of forest paths, giving the impression of a varied and uncorrupted countryside. Nothing was left to chance. "A park is a single work of art," Olmsted once wrote. "Every foot of the park's surface, every tree and bush, as well as every arch, roadway and walk has been fixed where it is with a purpose." Because of its curving pathways and well-defined spaces, the park unfolds to the visitor in a series of well-executed scenes. Buildings and other manmade features were kept to a minimum. The pathways and roads that carry pedestrians, horses, and vehicles circle the park in tidy, separate orbits. In a prescient move, the designers sunk below the surface of the park the four transverse roads that today accommodate Manhattan's cross-town traffic, thus preserving the pastoral effect.

The construction of Central Park was a project of awesome dimensions that took twenty years to complete. To create the park they had so carefully imagined, Olmsted and Vaux were forced to import million of trees and shrubs, and five hundred thousand cubic feet of topsoil to support them, all in horse-drawn carts. Thirty-six bridges and archways went up in the park. Lakes and ponds with artificial bottoms replaced the swamps. In 1863 the northern border was extended to 110th Street, bringing the park to its modern dimensions. Central Park was finally completed in 1873. Since then the most significant changes have come during the long tenure of Park Commissioner Robert Moses (1934–60), who saw the park as a space for sport and recreation in addition to the contemplation of nature. Under Moses, and over the objections of the strict conservationists, the park gained playgrounds, jungle gyms, sculptures, baseball diamonds, handball courts, and the Wollman skating rink.

Today visitors entering the park at its southeast corner at 59th Street and Fifth Avenue immediately leave behind the Trump Building and the Plaza Hotel and encounter the refreshing sight of the Pond, a tranquil body of water set among gentle lawns directly beneath the skyscrapers of midtown Manhattan. A tree-lined road called the East Drive runs north and west into the interior of the park, passing the Wollman Rink and the Central Park Zoo with its famous sea lion pool. At about 66th Street the Drive meets the Sheep Meadow, a fifteen-acre meadow that runs clear to the western edge of the park, opening stunning views of the New York skyline. Originally designed as a military parade ground, the meadow was in fact home to a flock of sheep and a shepherd until 1934. When the sheep weren't trimming the lawn they were housed in a Victorian building now occupied by the upscale restaurant Tavern on the Green. Today the meadow is home to sunbathers, picnickers, and kite flyers.

North of the Sheep Meadow lies the Lake, a 22-acre body of water created from a large swamp that twists and turns across the park between 71st and 79th Streets. Rowboats dot the lake in summer, just as they did in the nineteenth century. Behind the Lake looms the Ramble, a minutely planned miniature forest with winding paths, rustic shelters, quiet glades, and an artificial brook called the Gill. Because New York falls along the routes taken by migrating birds, the 37-acre Ramble is home to an amazing variety of birds, especially in spring and fall; 230 different species have been identified here.

Across from the Sheep Meadow, on the eastern side of the park, is the Mall: a stately promenade designed to provide a more formal element and a setting for strolling and socializing. Forty feet wide, the Mall is flanked by deep rows of American elms, a favorite tree of the designers. The southern end of the Mall, called the Literary Walk, features

A view of the tree-lined Mall. (© The Central Park Conservancy)

statues of famous writers and nonwriters alike. At the northern end is the Bethesda Terrace, one of the architectural centerpieces of the park. The grand terrace overlooks the "Angel of the Waters" fountain, capped with a beautiful bronze angel blessing the water, and just beyond it the Lake and the thick woods of the Ramble.

The middle section of the park, between 79th Street and 96th Street, is dominated by two main features, the Great Lawn and the Reservoir. To many, Central Park is synonymous with the Great Lawn, an egg-shaped lawn of thirteen acres in the heart of the park that can host eight baseball games or an outdoor opera with ease. In fact, the lawn was not included in Olmsted and Vaux's original design; at the time the site was given over to an enormous rectangular receiving reservoir for water piped down from the Catskills. In 1931, after the reservoir had become redundant, it was filled in to create the Great Lawn. Overlooking the Great Lawn from the south, perched on a huge outcropping of Manhattan Schist, is Belvedere Castle, a whimsical miniature castle designed by Olmsted and Vaux and built in 1872. Since 1919 the U.S. Weather Bureau has collected weather data from instruments mounted on the roof of the building. Looking north from the battlements of Belvedere Castle it is hard to miss the Reservoir, which fills the park like a jellyfish with its gently irregular shape. The 106-acre manmade lake once supplied Manhattan with fresh water. Today it is better known for the 1.6-mile running track that rings it, providing the city's joggers with an ever-turning view of the skyline. In 1994 the Reservoir was renamed the Jacqueline Kennedy Onassis Reservoir in memory of the former First Lady, who used to jog along its track.

In 1858 the terrain north of 96th Street was far more rugged and wild than the southern sections, and Olmsted and Vaux opted to retain the original character of the landscape. Today the northern end of the park is home to the North Woods, ninety acres of forest, steep hillsides, rocky bluffs, and tumbling water. Running down the center of the woods is one of the parks lesser-known treasures, the Ravine, a dramatic wooded stream valley. The stream is fed by the Pool, an idyllic pond on the western edge of the park near 102nd Street, which itself is fed by a 48-inch pipe camouflaged as a grotto. Water spirited up from the Reservoir flows through the Pool, over a waterfall, and enters the Ravine under a stone arch. On its journey north the stream tumbles over five waterfalls in total—all of them manmade—under a canopy of oak, elm, hickory, maple, and ash. At the northern end of the Ravine is the Huddlestone Arch, a massive stone arch built with boulders found near the site. As the Romans once did, Olmsted's builders fashioned the arch without mortar; the stones "huddle" together under the force of gravity. The stream empties into the Harlem Meer, an 11-acre lake with a miniature sand beach that is home to cormorants, swans, and grebes, all within sight of the cars on Duke Ellington Circle at the upper edge of the park.

Olmsted went on to design some of the greatest public landscapes in the United States, including Prospect Park in Brooklyn, park systems in Buffalo and Boston, the U.S. Capitol grounds, and the grounds for the 1893 World's Columbian Exposition in Chicago. Nonetheless, his name will always be linked with his first design. And whether or not they know his name, the millions who think of Central Park as their private backyard garden will always send him thanks.

 For more on the park, see www.centralparknyc.org. For more on Frederick Law Olmsted, see www.olmsted.org and www.olmstedsociety.org.

6

SUPPORTS, CLIMBERS, AND HEDGES

 Cicero's Trellis

Rather more is known about this Roman statesman who introduced the trellis than about the origins of the trellis itself. It is believed to have come into being as an agricultural device around 2000 B.C. when the Hittites and Egyptians observed that vines grew more vigorously if supported. When they started using the criss-cross structures for viticultural purposes, the Egyptians found they could also be used for marking out individual properties as perimeter fences. Another advantage was that they were light and fairly simple to construct. The Persians probably got to hear about this because they were reputed to have devised "diamond-mesh panels."

Full use of the trellis was certainly made by Cicero and his circle in the first century B.C. when decorative gardening began. Born in 106 B.C., Marcus Tullius Cicero was perhaps the greatest of Roman orators. Established as a prominent lawyer in 70 B.C. he was elected Consul in 63 B.C. His speeches and rhetorical works became models of Latin prose, and his letters provided a vivid picture of contemporary

Roman life. Stylish and persuasive, he was at the center of a circle which was influential in forming and developing decorative Roman gardens. The trellis not only became a favorite garden feature, it was also adapted for other uses. It was applied to the arbors and colonnaded pergolas which Marcus Porcius Cato had made popular with his book *De Re Rustica*, and which Caius Martius developed in the form of *romara tonsils* ("clipped arbors"). Cicero's writings also indicated the appearance of a specialist gardener called a *topiarius*, whose job it was to tend the ivy on trellises and pergolas. The extent to which the trellis developed into an

Cicero's trellis ideas were revived in Renaissance gardens, such as this one, where trellises were used for balustrading. (*The Gardeners' Labyrinth*)

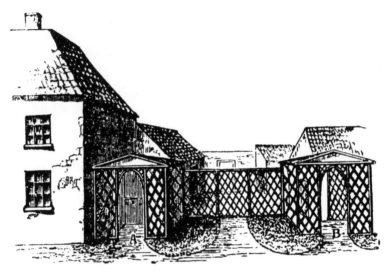

In Victorian times it came to be realized that the trellis was useful for hiding ugly architectural features. (Shirley Hibberd, *Rustic Adornments*)

elegant art form can be seen in the frescoes on the walls of Pompeii (and surviving because they were buried when that city was destroyed in 79 A.D.).

By then Cicero himself was no more. In his political capacity, after the assassination of Julius Caesar in 44 B.C., he supported Octavian (the future Emperor Augustus) and robustly condemned Mark Antony in a number of speeches known as the *Philippics*, for which he was arrested and, in 43 B.C., executed. Yet the development of the trellis appeared to be unstoppable. The next milestone was Emperor Hadrian's construction of a villa whose garden walls ingeniously combined trellises with *trompe-l'oeil* illustrations (which Cicero and his circle had also advocated) to make the lines of the trellis resemble those of the masonry. Eventually, in the eighteenth century, it reached the ultimate point of its development when it was used in France to construct vast pavilions. These were depicted in Batty Langley's *New Principles of Gardening.*

During Victorian times the trellis in its basic form enjoyed something of a revival as a means of extending growing space, partly through the enthusiasm of the populist writer James Shirley Hibberd, who championed its cause. Astonishingly, the simple trellis so widespread in gardens now is very similar to the type Cicero would have known.

For further information on Cicero and the trellis, see
www.arlisna.org/reports/9808 and www.users.ox.ac.uk.

🌿 Christopher Columbus and the Wigwam Frame

European gardeners have Christopher Columbus to thank for the introduction of the wigwam, a versatile (especially for growing vegetables) and much underused horticultural structure.

Born at Genoa in 1451, the son of a cloth weaver, Christopher Columbus worked in the same trade as his father for a while, until deciding on a career at sea. While still in his teens he went on a voyage to the Levant; he visited England in 1477, and may even have gone to Iceland. In 1479 he

The wigwam plant support idea was brought from the West Indies by Columbus.

So versatile were wigwam structures that in Scotland they were even used for drying corn. (*The Gardener's Magazine*, August 1828)

settled in Lisbon and married Filipa Perestrello. Believing he could reach the Orient by sailing westward he developed plans for an expedition, and spent seven years struggling to gain royal patronage to finance the voyage. Finally, on August 3, 1492, having gained the support of Ferdinand and Isabella of Spain, he set sail.

On Friday, October 12, land was sighted. Almost certainly it was one of the Bahama Islands, and Columbus named it San Salvador. Moving along the coast, he landed again on October 17. This time he got his first glimpse of Native American wigwams. They consisted of a roughly conical framework of poles stuck into the ground in a circle and converging to a point above the center where they were tied together; the frame was then covered with sewn bark, matting, woven rushes, or tanned hides. In his journal Columbus wrote: "Their houses are all built in the shape of tents, with very high chimneys." The crew who entered such dwellings reported that they were clean and neat. Returning to Spain, Columbus took with him some gold, tobacco, a hammock, a canoe, birds, and animals—and also six natives of the West Indies. He was received with the highest honor by the royal court; hailed as an admiral of the sea, he was created a grandee of Spain.

Descriptions of his sensational findings were initially disseminated in the *Condice Diplomatico Columbo-Americano* (commonly known as the "Book of Privileges"), produced by Columbus's son Fernando in 1502. Columbus himself died four years later. Nicholas Monardes then published a more

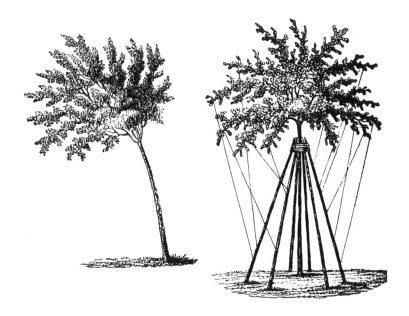

readable chronicle in 1569, entitled *Joyfull Newes out of the Newe Founde Worlde*. All the while the term wigwam, an English corruption of *wekou-om-ut*, meaning "in this house," was becoming more widely known.

Unlike the pyramid structures that tended to be used in smart formal gardens as devices for plants to grow up, wigwams always enjoyed a rather rustic quality. They came to be much used in

Wigwam supports were used for a variety of tasks including (above) supporting fruit trees (Loudon, *The Villa Gardener*) and (right) training hops in the French Vosges region (*The Gardener's Magazine*, February 1832). They could also be used to protect trees in the open from frost.

Victorian times for a variety of purposes. In agriculture they served as stands for drying corn and as supports of frail trees, while in the French Vosges in 1828 they were found to be ideal for growing hops.

 For extracts from Columbus's journal, see www.fordham.edu/halsall; also see www.columbus.gl.iit.edu and www.indians.org.

William Chambers's Bamboo

Bamboo is now widely used in western gardens for a multitude of purposes, such as constructing garden buildings or as supports for growing plants. Various types of bamboo are also grown as specimen plants. Sir William Chambers did not personally bring bamboo to Europe but he was the first influential architect to extol the virtues of such Asian garden accouterments in his lifetime—and he was publicly ridiculed for doing so!

William's Scottish grandfather was a wealthy merchant who had supplied the armies of Charles XII of Sweden with supplies and money, and had suffered by receiving base coin issued by that infamously late-paying monarch. His father stayed in Sweden for many years recovering the loan, then returned to England in 1728, bringing with him the infant William who had been born in Gothenburg in 1726. On completing his schooling at Ripon in Yorkshire, William Chambers returned to Sweden to work for the Swedish East India Company. At the age of sixteen he became a cargo supervisor on one of their ships which took him to the coastal city of Canton.

Here he was fascinated by the glimpses of a romantic civilization so completely different from his own, and for the next two years he spent a considerable part of his time ashore with his sketchbooks. His hobby was not without peril, for, as he later wrote, it was a "matter of great difficulty to measure any public work in China with accuracy, because the populace are very troublesome to strangers, throwing stones and offering other insults." He persevered nevertheless, and was evidently

Sir William Chambers (1726–96). (*Journal of Horticulture*, September 1903)

considered trustworthy enough to be granted privileged access to a number of private homes and gardens where Europeans had never before been admitted. It was in those gardens, which were mostly quite confined, that Chambers saw for the first time the horticultural uses to which *bambusa*—as bamboo was then called—could be put.

For two years, on and off, he made sketches and drawings of such gardens, buildings, and other forms of Chinese design. By 1744 he had saved enough money to leave his job with the purpose of studying and training to become an architect. He was taking a risky step, especially as his brother John, who had similarly embarked on a life at sea, remained in the East India trade and amassed a sizable personal fortune. But William had enough faith in his own abilities to spend a long time traveling in Europe, noting the work of classical and Renaissance architects. He enrolled at the *Ecole des Arts* in Paris, then studied in Rome with the sculptor Joseph Wilton, whose beautiful daughter he charmed with his earthy humor into being his mistress.

Later, he married her. Returning to England in 1755, the Chamberses settled in London's Poland Street, where William began his career as an architect.

William's opportunity to become a gardening pioneer came astonishingly quickly. Within two years of setting up shop he was brought to the attention of the Earl of Bute. This nobleman was on especially intimate terms with the Dowager Princess of Wales and he recommended Chambers to her as a suitable person to instruct her son, later George III, in drawing and architecture. This appointment brought Chambers the royal favor which he enjoyed for the rest of his life and undoubtedly helped to place him at the head of his profession.

A wave of interest in China and Chinese artifacts was sweeping across Europe at that time. Chinoiserie furniture based on Chinese designs began to appear which was extravagant to the point of absurdity, as were the garden designs in William and John Halfpenny's *Rural Architecture in the Chinese Taste* (1750). As one of the very few trained architects who had been to China, Chambers found himself well placed to be a leading light of this movement. Careful to provide accurate depictions, rather than be carried away by an excess of enthusiasm for all things Chinese, in 1757 Chambers published *Designs for Chinese Buildings*, the illustrations in which were closely derived from the sketches he had made in Canton. Plate IX of that book effectively provided the western world with the earliest image of bamboo in a small courtyard garden. There it was used to construct an arbor, and was also growing as a decorative plant. The image showed a typical well-to-do private house which he described as follows: "Every apartment has before it a court with an artificial rock placed there in which grows some bambusa." That spring he enthused about the use of such places in an article for *The Gentleman's Magazine* called "The Art of Laying out Gardens among the Chinese," in which he further described what he had seen in Canton.

Chambers had no wish to see the Chinese style adopted on a large scale in Europe, but he reckoned it might have a place in extensive parks and gardens where a variety of scenes was required. His greatest chance to put that vision into practice came when he was invited by Princess Augusta to embellish her "gardens" at Kew. It was here that he made his reputation with the grounds he laid out, and the two dozen ornamental buildings—including Asian bridges, retreats, and temples—which were erected at Kew between 1757 and 1762. His book *Designs for Chinese Buildings at Kew* (1763), which gave details of the structures he had recreated there, contained a picture of an elongated menagerie, the slats of which in China would almost certainly have been made of bamboo.

When the Prince of Wales ascended to the throne, Chambers became the royal architect. In 1771 the king permitted him to assume a knighthood on account of an equivalent honor—the Polar Star—which had been bestowed on him in Sweden. Chambers was now a senior establishment figure and sought to use his position to record his strong contempt for the prevailing, and largely unchallenged, taste for bland landscape gardening as advocated and perpetuated by "Capability" Brown. His venomous attitude toward Brown may have been exacerbated by the fact that Clive of India had chosen Brown, not Chambers, to design Claremont, a mansion in Surrey, much to the latter's chagrin.

Chambers went on the attack in 1772, when he published *A Dissertation on Oriental Gardening*, which gave a sarcastic and fantastically exaggerated description of Chinese gardens. This stirred up considerable resentment toward Chambers among Brown's followers. In early 1773 the poet William Mason (author of *The English Garden*) and Horace Walpole anonymously brought out a satirical riposte, *An Heroic Epistle to Sir William Chambers*, lampooning the architect. Hugely popular, the *Epistle* went into fourteen editions. Part of the reason for its success was unquestionably due to the fact that it mixed politics with gardening, since any attacks on the unpopular royal party (with which Chambers was identified) were at that time sure of a favorable reception.

This devastatingly mocking counterattack not only made Chambers look foolish, it also neutralized interest in Chinese gardening ideas in Britain. As a result, it went largely unnoticed that in the *Dissertation* Chambers praised the Chinese for making gardening a distinct profession, attained by extensive study. Significantly the book had listed many plants and shrubs grown in China, a number of which were still completely unknown in Britain, such as the bamboo. Indeed, on the continent, where the *Heroic Epistle* made no impact, Chambers's *Dissertation on Oriental Gardening* and *Designs for Chinese Buildings* both exerted a strong influence. Translated into French, they stimulated a high degree of interest in Chinese gardens and gave added authority to the term *anglo-chinois*. Strangely enough, as a garden pioneer Chambers was appreciated almost everywhere except in his own country.

Stung by the adverse reaction to his stance in Britain—he had effectively been laughed out of court—Chambers put aside his interest in gardens and for the rest of his working life concentrated virtually all his professional attention on architecture. It was a sensible move because it enabled him to accumulate a considerable fortune. From 1775 onward he was engaged in remodeling Somerset House, while his magisterial *Treatise of Civil Architecture* became a standard textbook for students. Regardless of the political lampooning, Chambers retained a cheerful and convivial disposition, and his wide circle of influential friends included Dr. Samuel Johnson, Oliver Goldsmith, David Garrick, and Sir Joshua Reynolds. The perfect gentleman, he always dealt fairly and honestly with craftsmen and artisans. Everyone would be invited to the fêtes and dances at Wilderness House, his great Palladian villa in Twickenham, where he lived a grandee's existence. He died from a grave illness, exacerbated by acute asthma, on March 8, 1796, and was buried at Poets' Corner in Westminster Abbey.

There were several lasting monuments. In garden terms the most spectacular was, and remains, the ten-story Chinese pagoda at Kew, which stands at the heart of several vistas. Bamboo, the introduction of which Chambers had done so much to encourage, is usually considered not to have been brought from the East into Europe until 1827 (in the form of the sturdy black bamboo). Yet by the early 1790s tall bamboo bushes were flourishing in the garden of the senior statesman Lord Grenville, at Dropmore Lodge near Windsor. Knowing that, Chambers would have died a contented man.

 Chambers's pagoda and many different types of bamboo can be seen at the Royal Botanic Gardens, Kew, Richmond, Surrey TW9 3AE; www.rbgkew.org.uk. France's leading bamboo center is Bambouseraie Prafrance, Generargues, 30140 Anduze; www.bambouseraie.fr. Useful link: www.bamboobridge.com.

🌿 Christopher Leyland's Monster Hedge Tree

American gardeners in the southeastern states are tremendously enthusiastic about the vigorous evergreen Leyland cypress tree, which has won popularity there because of its Christmas tree looks, although it has become notorious in Britain as a nuisance hedge. It may come as a surprise that it was originally discovered quite by chance by a heroic nautical pioneer.

Christopher John Leyland was born on September 19, 1849, the first child of John Naylor, who was head of an immensely wealthy firm of Liverpool bankers. The Naylor-Leyland family was accustomed to living on an epic scale. During the summer months they would tour the Mediterranean and North African coast in their steam yacht. In 1849 John Naylor had acquired Leighton Hall, a 3,916-acre country estate near Welshpool in mid-Wales. Wishing to satisfy his long-held desire to be a civil engineer, he set about extensively transforming the estate into a state-of-the-art mechanized model farm. By 1855 he had established a redwood pinetum on the slopes of the adjoining Long Mountain, stocked the park with kangaroos and bison,

Christopher Leyland, the discoverer of the infamous Leyland cypress, was a wealthy landowner and a national maritime hero. (*North News & Pictures*)

restyled the mansion with a 328-foot observation tower, constructed gigantic new farm buildings, and equipped the estate with a private funicular mountain railway, powered by newfangled water-turbines, fed by the mountain streams. Water power also drove the estate's sawmills. Assembled on a scale unparalleled elsewhere in Britain, the estate contained some of the most ingenious feats of engineering of the Victorian age.

It was here that Christopher Leyland spent his early years. In 1862 he entered the Royal Navy, and served for ten years before retiring as a sub-lieutenant. He then took an investment position in the family bank. Possessing some unusual knowledge about turbines, largely derived from the ones he had seen on the estate, he began giving financial advice to the steam-turbine innovator Charles Parsons. As his father's heir apparent Christopher also took a practical interest in the forestry activities at Leighton Hall.

All this while a rare Nootka cypress (*Chamaecyparis nookatensis*) had been growing among the specimen trees in the Leighton Hall pinetum. This robust cedar was commonly found below the snow-lines of glacial slopes across America, from Oregon to Alaska. Just to windward of the Nootka cypress and also flourishing in the glade was a Monterey cypress (*Cupressus macrocarpa*). This rare tree was only to be seen growing in the wild on a mile-long stretch of low cliffs near Monterey in California. These two trees belonged to quite different sections of the genus and would never normally have shared the same habitat. Natural cross-breeding was a rare occurrence in plants, particularly in conifers, but in 1888 Christopher Leyland gathered and sowed some seeds from a cone from the Nootka. With his fine eye for detail he glanced at the fledgling seedlings and noticed that six of them differed slightly from the rest. Had he destroyed them the history of trees would have been much the poorer. But, such was nature's fortune (some neighbors might say misfortune), Leyland had a tolerant heart and chose to spare them.

Christopher's life was now nearing a transformation stage. He had inherited Leighton Hall from his father in 1889, but in the following year his wife of fifteen years died. In 1891 a great-uncle also died, and bequeathed him another huge estate, Haggerston Castle near Berwick-upon-Tweed in Northumberland. By this time Leyland had become the main investment-finder for Parsons's group of turbine companies (with whom he held several directorships). Furthermore, he had already been efficiently managing Haggerston for his great-uncle. Resolved to make a fresh start he sold Leighton Hall to his brother and moved to Northumberland, where he made Haggerston his personal residence and spent the next few years extensively rebuilding it. With him he took the six young trees, which he planted in various parts of the castle garden.

There, as they grew, it became evident that they were the result of a natural cross. Quietly delighted by the discovery, but without telling anyone beyond the confines of his family, Christopher named the creation Leyland cypress (*Cupressocyparis leylandii*). He allowed cuttings to be taken from the trees and distributed to various gardens on the estate. But at Leighton Hall in 1911 there was another extraordinary occurrence. Leyland's nephew Captain J. M. Naylor picked a cone from a Monterey cypress growing 164 feet away from a Nootka. Two seedlings of this batch appeared slightly different from the rest and, mindful of Uncle Christopher's 1888 discovery, Captain Naylor planted them. Fourteen years later he noted that they had grown to 33 feet in height, whereas a Monterey cypress from the same seed-bed had reached a mere 23 feet. Evidently the Leyland cypress not only grew faster than either of its parents, it was also promising to grow far taller.

The outside world only learned about the existence of the burgeoning monster trees at Leighton Hall and Haggerston when a tree specialist who had visited Leighton Hall sent a sample of the previously unseen foliage to the Botanical Gardens at Kew in July 1925. News of the discovery, which was hailed as "one of the most important tree introductions of recent times," was broken to the public in the *Kew Bulletin* of autumn 1926. Cuttings were taken by Kew Gardens to produce stock-plants from which were eventually derived eight varieties of Leyland cypress, with names such as "Naylor's Blue," "Haggerston Grey," and "Leighton Green." From these were raised clones that were distributed around the world.

The Leyland cypress seemed to possess all the virtues of the Monterey cypress, but none of its vices. Like the Monterey, it was very resistant to sea-winds. Whereas the Monterey was restricted to areas with a milder climate, the Leyland cypress inherited from its other parent the ability to resist all but the worst winters without damage. It made a fine specimen tree, but was also a reliable and fast-growing shelter-belt tree. Moreover, people hoped that it would also make an excellent hedge plant—and sure enough, its ability to withstand such restriction was shown by the flourishing dwarf hedges in the arboretum nursery at Kew. Planted in 1947, the hedges were kept down by mechanical trimming to a height of 5 feet—and reportedly showed no resentment of such drastic treatment.

For many gardeners the prospect of trimming the Leyland cypress was less important than its qualities as a fast-growing hedge that could thrive in almost any soil. Nurserymen pushed it as a sales item because they could grow it easily and therefore sell it cheaply and in enormous numbers. Then along came the supermarkets and other large retail outlets, which also realized how much profit could be made out of the Leyland cypress. In fact, because of its rapid growth rate, it was singularly unsuited for the purpose of hedging in all but the largest gardens. Certainly it could produce an effective hedge in three to five years—but few people who bought the Leyland understood that it could grow at the same rate for years! Indeed, the Leyland cypress could reach heights of 138 feet!

Owners soon found that controlling the tree was an onerous annual chore. If the trimming of the sides was neglected for a short while, the rapid growth meant it would be necessary to cut back into the old wood, leaving ugly dead sticks that would never produce leaves again. The dry heart of the tree was extremely flammable—if it caught fire it would burn like a torch—and hedges planted close to a house constituted a fire hazard. If allowed to grow naturally, the tree was apt to lose its lower branches in heavy snow, and that could result in tall but very scruffy hedges. It was even inhospitable to wildlife because it offered virtually nothing in terms of food. But the Leyland cypress's reputation as a nuisance "hedge-from-hell" only became apparent in the 1990s, with various celebrated legal wrangles between petty-minded neighbors. Disputes ranged from damage to houses caused by roots to the blocking of natural light by untrimmed growth.

Belgium, France, Germany, Holland, and Switzerland have all found it necessary to introduce hedge-nuisance legislation. The Leyland cypress was grown far and wide, and in New

Zealand and Australia it is used for wood products. Rooting cuttings arrived in the United States, through California, for the first time in 1941. In 1965 they found their way to South Carolina, where their potential for use as Christmas trees became apparent to the State Forestry Commission and private growers alike, and for that purpose they have been successfully grown in California, Louisiana, Mississippi, Alabama, Georgia, Florida, and the Carolinas.

Oblivious to the horticultural mayhem he was storing up for future generations, at Parsons, Christopher Leyland became a national hero as the financier and captain of the world's first steam-turbine ship *Turbinia*. Experimental models of her hull were tested by Leyland in the main lake at Haggerston. Parsons's revolutionary steam-turbines went on to power transatlantic liners such as the *Mauretania*. Leyland is also believed to have had a financial interest in the Moor Line, a steamship freight company. He lived out his later years as an eager member of the Royal Yacht Squadron, sometimes quietly taking cruises in the North Sea in his steam yacht *Tourmaline*. Or so it seemed.

In 1896 a Moroccan rebel leader called Dr. Abd-el-Kerim Bey secretly visited London to meet a financial syndicate and to plead for humanitarian support and weapons for use against the cruel French colonial occupying forces. A party of British mercenaries was subsequently recruited to covertly deliver a large quantity of weapons to Moorish freedom-fighters on the Moroccan coast near Mauritania. There is a possibility that Leyland himself was involved in this so-called "Tourmaline Expedition," particularly because the boat in which the arms were illegally shipped to North Africa was his 150-ton yacht!

As a magistrate and the chairman of the Berwick division of the Conservative Association, Leyland had good reason to

Turbinia's audacious 35-knot demonstration at the Spithead Naval Review in 1897, with Christopher Leyland at the controls. Millionaire Leyland was the financier and main captain of the *Turbinia*, the world's first steam-turbine vessel. (*Heaton Works Journal*, June 1935)

Fast transatlantic liners—with masts as tall as giant trees—were soon being powered by steam-turbines, as pioneered by Leyland's *Turbinia*, pictured here alongside the *Mauretania* in 1912. (*Heaton Works Journal*, June 1935)

leave few clues which newspapers could use to implicate him with the expedition. Furthermore, all the while this gun-running caper was going on, Parsons was seeking to sell the *Turbinia* concept to the French navy. (Indeed, Leyland was due to pilot the *Turbinia* along the Seine to display at the 1900 Paris Exhibition.) From the outset the mercenary expedition was a fiasco. The Foreign Office treacherously betrayed details of its activities to the Moroccan sultan, so that when the yacht reached the North African coast the crew's key personnel were captured and imprisoned. Leyland had wisely not appeared on board once the boat had left England.

Ashore Christopher Leyland cared greatly for the welfare of his numerous tenants and won a reputation as a humane and just landlord. In October 1926, within weeks of the *Kew Bulletin*'s revelation, he died near Haggerston Castle. He took with him the secret of how the parent trees of the Leyland cypress actually came to the Leighton Hall pinetum from North America. Now we will never know!

 Christopher Leyland's internationally famous steam-turbine demonstration ship *Turbinia* is on display at the Discovery Museum, Blandford Square, Newcastle upon Tyne NE1 4JA; also see the useful website www.birrcastle.com.

FERTILIZERS AND PEST CONTROLS

 Columella's Soil Test

Nowhere is soil more intensively worked than in gardens, but insofar as there had been any literature on the subject of soil it had been confined to agriculture—until Lucius Junius Moderatus Columella, that is. Unusual for an influential Roman, Columella was a native of Cadiz, Spain. In early adulthood he had been a tribune of the legion stationed in Syria. Neither military life nor a career in law appealed to him, and when he came into a substantial amount of capital—by unknown means—he chose to quit the army to become a farmer. Investing his fortune wisely in Italian land he was able to spend his leisure time writing at length on his favorite occupation. His scribblings benefited from his practical experience of agriculture. A stickler for high-quality produce, especially fruit, he experimented with fifteen varieties and many cultivars. He also made the first written mention of apricots.

Columella's eloquent and influential *tour de force* was *De re Rustica*, which he published in twelve volumes between 60 and 65 A.D. The longest and most comprehensive of all Roman agricultural manuals, it also dealt with the cultivation of gardens. Although, like many Romans, Columella was drawn by temperament to the land, he was detached enough to hold sound views on how it might best be cultivated. For Columella, there were no romantic spiritual notions of the earth. In his writing he scornfully mocked those fatalists—and even administrators of the state—who adhered to the widely held belief that poor crops and declining soil fertility were due to the soil being worn out and exhausted by overproduction. He questioned the conventional view that compared Mother Earth with a human mother: "For it is a sin to suppose that nature, endowed with perennial fertility by the creator of the universe, is affected by the barrenness as though with some disease." That and soil erosion, he insisted, were due to improper land management. Soil exhaustion was not related to the age of the earth but to the standards of agricultural practice. He argued that the most fertile soils, quite irrespective of age, were to be found on forest floors, owing to the accumulation of fallen leaves. Curiously, however, he never went so far as to encourage people to produce organic mulch for use on their farms and gardens.

Gardens were dealt with in two books of *De re Rustica* (the tenth and eleventh), in which Columella covered the layout of gardens, their water supply, the use of fertilizers, the plants to be grown, and details regarding their culture. He placed much emphasis on the importance of soil preparation, much of it based on sound common sense. Shrubs, for example, he claimed should be moved into soil similar to that in which they had been growing.

Columella insisted that color was a simple and vital indicator of soil quality. He advocated a "soil test" which involved the kneading of dampened soil by hand. If it "sticks to the fingers of the person holding it, in the manner of pitch, it is fertile." One recommended experiment involved digging a hole and then refilling it with the original soil. If, when trodden in, there was some left over, the soil was probably fertile, because poor soil would not refill the hole. Another method involved mixing soil and water, straining it, and tasting the result: a sour taste meant poor soil, a sweet taste meant good. "Sweet ground" would typically grow wild plants such as rushes, reeds, and brambles.

Many years later (in 1725) pearls of wisdom such as these were brought to the attention of modern horticulturists in Richard Bradley's translation of Columella's writings, *A Treatise on Ancient Husbandry and Gardening.* As for the Spaniard who had spent most of his life in Italy, little is known of him—not even his dates of birth and death.

For further information on Columella, see www.deer.r and www.gardenvisit.com.

Leonard Mascall, Grub Catcher

How to deal with slugs and snails is a perennial problem that has vexed gardeners for centuries. Until Leonard Mascall's time, if there were remedies at all, they were handed down almost as a matter of folklore. Mascall's significance is that he is believed to be the first person to have written down his miracle cure in a published book and thereby supplied it to a wider audience.

Slugs and snails are among the most bothersome pests in many gardens on both sides of the Atlantic. They attack both above and below the ground and eat the leaves *and* the roots. The most common species is the *hortensis* or garden snail, which, having started munching through a leaf or fruit, will not begin on another until the first is wholly eaten— such polite guests! The common brown snail, which causes the most problems in Californian gardens, was introduced from France during the 1850s for use as food. Both snails and slugs belong to the mollusk phylum and are similar in structure and biology, except that slugs lack the snail's external spiral shell. Of the slugs, arguably the most mischievous have long been the gray field slug, black slug, garden slug, and keel slug.

Leonard Mascall, who wrote about the advantages of picking off grubs by hand. (*Journal of Horticulture and Cottage Gardener*, 1875)

The name Mascall is probably a corruption of the Norman *Marescal*. The founder of the family probably came over with William the Conqueror, and members of the family became sheriffs of Sussex in the reigns of Richard I and John. Their chief residence was Plumpton Place at Lindfield in Sussex, an impressive stately hall with a drawbridge over the wide moat that surrounded it. When Leonard Mascall was born there in 1546 the mansion had already achieved historical recognition since it was here that de Montford marshaled his ranks for a battle against Henry III. Of Leonard's early life virtually nothing is known, although he is believed to have become Clerk of the Kitchen to Archbishop Parker, then Archbishop of Canterbury. Perhaps it was in that capacity that he traveled on the continent. He was undoubtedly acquainted with the French and Dutch languages. Furthermore his entrepreneurial verve and practical knowledge of fruit was such that he became responsible for planting the first pippin apple tree in England.

Mascall set out his advice on trapping slugs and snails in a book entitled *The Country-Man's Recreation or the Art of Planting, Grafting and Gardening*, published in 1640. Three sections dealt with "Planting and Grafting," "Hop-Gardens," and "The Expert Gardener, Containing Various Necessary and Rare Secrets Belonging to that Art." In the latter, he quaintly instructed:

> You must take heed of all manner of young trees, and especially of those grafts, the which many worms do damage and hurt in the time of Summer; those are the field snail which hurtteth also all other sorts of trees that be great, principally in the time the cuckoo doth fly, and betwixt April and mid-Summer, while they be tender.

Moving on to the crucial section, "How ye ought to take the said worms," he revealed his killer method:

> For to take them well, you must take heed and watch your young trees in the heat of the day and where you shall see any, put your hand softly underneath, without shaking the tree, for they will certainly fall when one thinks to take them; therefore so soon as ye can (that they not fall) take them quickly with your other hand.

Hand-picking slugs and snails, the pest eradication method advocated by Mascall, though primitive—and unpleasant— was effective. In subsequent centuries it developed into something of a fine art. In 1712 John James's *Theory and Practice of Gardening* enthused that snails could easily be taken by hand if looked for in the mornings and evenings, especially after rain. In *An Encyclopaedia of Agriculture* (1825) John Loudon noted a refined method of picking that was extensively practiced by market gardeners and others. Immediately after turnips had been sown, the ground was strewn with cabbage leaves. On these the slugs would "pasture," especially if the leaves were beginning to decay (which produced a sweetness), and they could be easily gathered every morning. By depositing a cabbage leaf every square foot or so, a whole field could soon be cleared by picking off slugs and removing the snail-laden leaves every twenty-four hours. Then in 1905, Robert Thompson's multi-volume *The Gardener's Assistant*, referring to cabbage leaves as *traps* for slugs and snails, insisted that "Lettuce leaves, as well as pieces of board loosely on the soil, would enable the cultivator to secure large numbers of the pests in the vicinity of valuable plants." Once caught, the pests could be destroyed by crushing underfoot, or thrown into salt or boiling water.

Mascall's hand-picking idea certainly did not preclude the development of alternative mechanisms for removing slugs and snails. They could be deterred by scattering caustic substances over them, or by "watering" them with diluted nicotine or bitter infusions such as vinegar or strong lime-water. Then there were abrasive dusts. Rings of coal- or wood-ash, coal, and charcoal-dust laid around valuable plants would save them from attack unless there were slugs already inside the ring. Copper wire or foil could also be used to encircle precious plants. More generally, slugs and snails could be caught with beer-baited traps, such as deep, smooth-sided jars buried at soil level, in which they would drown. Another innovation came in the form of chemical pellets, although they were not always effective and had the disadvantage of being harmful to children and pets.

In 1596 Mascall wrote *On the Government of Cattel*, a book on animal husbandry. A few years later he was appointed to

The commonest grubs are the garden snail *H. hortensis*, and the slugs *Limax ater* and *Limax agrestis*. (Loudon, *An Encyclopaedia of Agriculture*, 1825)

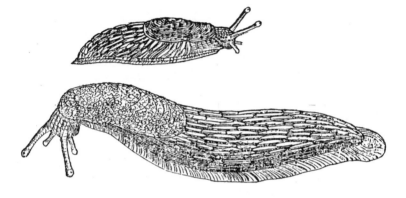

the office of royal farrier—a curious appointment which appears not to have required him to have any skill in the craft of shoeing or leeching horses. It is likely to have been bestowed upon him because, like his royal master, he was a scholar and believer in witches (in the book he explained how to tell "the difference between a horse bewitched, and other soreness"). This gives cause to wonder if he might have regarded slugs and snails as the devil's creatures.

He wrote several other books, including, bizarrely, a tome on removing stains from clothing. He was the earliest Englishman to write about poultry-keeping, and in 1590 he wrote *A Book of Fishing with Hooke and Line*, which seemed to confirm the legend that he had introduced carp into England from the Danube, placing them in the deep moat at Plumpton Place. He died in 1605, and it was only a great many years later that doubts began to arise about the origins of his writings. In early 1851, while searching the catalog of the British Museum for Mascall's 1581 book on poultry, a writer for the horticultural journal *The Cottage Gardener* noted a discrepancy which seemed to indicate that it might have been a translation of a book by a French author. This had the effect of focusing attention on the originality of his other works. Closer inspection showed that he had freely acknowledged his use of translations. The episode dented Mascall's reputation, and undoubtedly lent him something of an air of mystery, although it did not devalue the practical value of his observations.

Indeed, Mascall's effective organic approach to ridding gardens of slugs and snails now has an increasing contemporary relevance. He could have added that natural predators can do their bit. Song-thrushes and blackbirds are able to destroy large numbers of slugs and snails, and should be encouraged in and about the garden. Another method is to keep a pet which loves eating slugs—a pet mole!

For further information on garden pests, see
www.slugcontrol.iacr.ac.uk; and www.oxalis.co.uk/slug.

Robert Sharrock's Mole Trap

In 1885 the *New York Tribune* excitedly reported that a gardener, whom it omitted to name, had recently devised a means of trapping moles which he wanted to make known to the public for their free use. This mole-trap was essentially a large flowerpot—an old tin pail was claimed to answer the purpose excellently—sunk beneath the ground upon a level with the floor of the mole's tunnel. A flat piece of board was laid over the top, and earth heaped upon it so as to exclude the light completely. His success lay in the perfect simplicity of the device. The moles, seeing or feeling nothing with the highly sensitive "feelers" upon their snouts, would run very readily into the trap, from which there was no escape. Every fresh arrival simply added to the company, for no resetting was required, and there was no disturbance of the ground to excite suspicion. Doubtless the movement of the captured moles would attract others to their ruin.

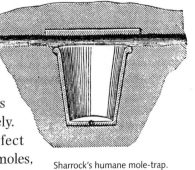

Sharrock's humane mole-trap. Scurrying along a tunnel, the mole would fall into the pot. (Mrs. Beeton, *The Book of Garden Management*, 1872)

In fact this ingeniously simple contrivance was not a new invention at all. It had been described by Robert Sharrock as long ago as 1694. Born in 1630, Sharrock became Arch-deacon of Winchester Cathedral. In those days Winchester was a significant urban center, and presumably Sharrock made use of his leisure time to escape into the countryside to gather natural history data and anecdotes. He was a member of the Royal Society and was well regarded by his contemporaries. He published his views and findings in *The History of Propagation and Improvement of Vegetables by the Concurrence of Art and Nature* (1660), a book which went through three editions, and then produced *An Improvement to the Art of Gardening* (1694), in which he set out his mole-trap ideas.

A mole was instrumental in causing the death of William III, who fell from his horse when it tripped over a molehill at Hampton Court. Although toasted as "the little gentleman in the black velvet coat" by recalcitrant Jacobites, most agriculturists and gardeners on both sides of the Atlantic still regard the mole as a pest. This is despite the fact that moles can have a beneficial influence in horticultural settings: not only do they eat pests such as slugs and wireworms, which are injurious to many plants, but also their tunnels help to drain and aerate heavy soils.

Even in Robert Sharrock's time moles inhabited most parts of the world. There are about thirty species, ranging from 4 to 9 inches in length. They spend almost their entire lives underground and their bodies are designed accordingly. The head has a tapering snout, the external ears are small or absent, the eyes are minute and imperfect—some species even have the eye covered with skin—and the neck is short and muscular. The bones of the shoulder and arm are squat and strong and their muscles are enlarged and powerful. Also distinctive are the stubby front feet which are held in a hands-over-the-head position with the palms facing outward. All five fingers are present on the forefeet and are well developed, each with a strong claw. With these powerful forefeet moles can literally swim through the soil, and are capable of tunneling as far as 984 feet per day. However, the real problems for gardeners start when the soil is particularly compact and the mole has to dispose of the surplus earth. It is shoveled up to the surface at intervals, creating the familiar molehills.

The mole actually lives in a permanent, larger network of tunnels, termed a fortress. There is in fact no symmetry or planning about it, and the so-called galleries are simply the tunnels through which all the earth excavated to form the nest chamber is shoved up to the surface to form a large molehill. The nest chamber itself is a spherical cavity some 10 inches in diameter and lined with dry grass. It appears that each mole, male or female, has its own fortress, connected with a network of tunnels along which it goes foraging. Each mole builds a bolt run that leads downward from the nest chamber and then turns away from the fortress independently of the runs that are in frequent use. The network of a male mole usually consists of a long main tunnel with many branches, while that of a female is inclined to be more reticulate and have no well-defined highway.

Earthworms are the most important component of their diet (a mole weighing 2½ ounces needs to munch through 1½ ounces of earthworms per day), and the moles patrol their shallow tunnels just below the surface during the day in search of worms and other grubs which have fallen in.

The traditional method of catching moles involved a noose which was buried on a board, and sprung with a hazel branch. In America the most popular device came to be the spring-loaded harpoon spike; elsewhere, especially in Europe, gardeners favored crushing in a scissors trap (above) or a barrel trap (opposite). (*The Gardener's Magazine*, June 1832)

This is when the mole is most vulnerable. In Sharrock's time the commonest method of catching moles involved a trap: A noose-like loop was attached to a wooden board buried in the run; the noose was held with a catch and sprung on a bent hazel stick. Other means came to be tried at various times, such as flooding the tunnels with water (although most moles could swim). Foul-smelling items were often inserted into the tunnels, such as old fish heads, red herrings, garlic, leeks, and moth repellent. More drastic measures involved lighted rags soaked in paraffin and even worms baited with strychnine. Spring-loaded mechanical traps came to be by far the most popular means of catching moles. There were three main types: the scissors and barrel traps both killed by crushing, while the harpoon trap

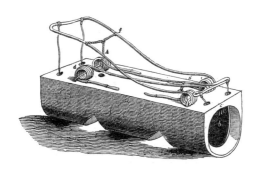

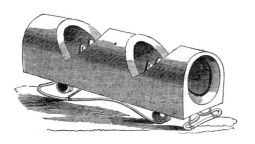

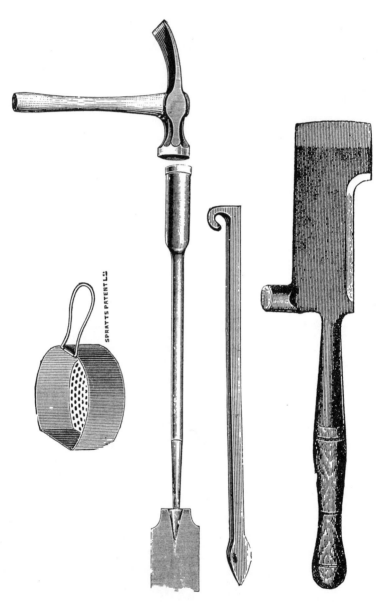

Various tools were devised for digging in the mole-catching contraptions. (W. Carnegie, *How to Trap and Snare, c.* 1900)

(especially popular in North America) impaled the hapless mole on spikes. The disadvantage of most of these grisly methods was that in any domestic garden they could be dangerous to children and pets.

Sharrock's remedy, which he outlined in *An Improvement to the Art of Gardening*, derived from the advice of a local farmer near Winchester. He claimed:

> Mr. Bligh relates that one spring, about March, one mole-catcher, in about 10 days, in an area of ninety acres, captured 3 barrels-full of old and young moles. By burying the barrels just below the surface near the moles' nest some of the moles would fall in, then hearing the noise the old would come and look for the young.

It was a brilliantly simple, relatively "humane," and very efficient means by which to catch moles—and one which would be acceptable to today's organic gardeners.

For more information on moles, a useful website is www.moletunnel.net.

William Forsyth's Tree Plaster

As King George III's head gardener, William Forsyth should have gone down in history for his achievements as the creator of Europe's earliest rock garden and as a founder of the Horticultural Society; instead his good name was sullied by scandalous allegations that his discovery of a miraculous tree plaster was a fraud.

William Forsyth was born at Old Heldrum in Aberdeenshire in 1737. He is believed to have been an apprentice in the gardens of Lord Aberdeen at Haddow House, although beyond that nothing much is known of his early life. Like many other Scotsmen William soon went south and by 1763 he was working at the Apothecaries Garden at Chelsea with the celebrated gardener-botanist Philip Miller, author of *The Gardener's Dictionary*. He was also employed further along the Thames at the Duke of Northumberland's Syon House garden. It was in 1771 that his chance for real advancement came. Complaints were made about Miller's conduct and the old veteran, aged nearly eighty, resigned; luckily for Forsyth, the Grand Committee appointed him as Miller's successor.

Changes soon began to take place at Chelsea. In 1773 Superintendent Forsyth oversaw the building of England's first rock garden. Ingeniously he allowed it to be constructed with 40 tons of demolished stonework from the Tower of London, plus various flints and gravels. To enhance the effect, it was all mixed with a large quantity of volcanic lava brought back as ballast from Reykjavik in Iceland by Sir Joseph Banks's ship *St. Lawrence*.

In 1784 Forsyth was prompted to take a closer interest in fruit cultivation when he was appointed head gardener to George III at St. James's and Kensington Palaces, a post that required him to provide a ready supply of quality fruit for the royal table. Many of the existing fruit trees were showing signs of tiredness;

William Forsyth. (*Horticultural Who was Who*, 1948)

there was gumming in the stone fruits, and "canker" (a dangerous fungal infection) was becoming rife in the apples. Nevertheless, because he had to maintain a supply of fruit, Forsyth hesitated to replace the old trees with healthy young ones. In order to save the old trees he lopped off many large diseased branches, and in some cases had them "headed back" (virtually pruned to the ground). All this involved the making of large wounds. In order to protect the bare patches until a callus had grown, he covered them with a mixture of cow-dung, wood-ashes, and sand, beaten into the consistency of ceiling plaster. The same concoction was used when he had occasion to remove large or damaged branches from elms and other trees.

In 1789 England went to war against Napoleon, and the provision of oak trees for building warships and merchant vessels was increasingly a matter of national importance. It was then, while investigating the Crown Estates, that the commissioners of the Land Revenue Office suddenly made an alarming discovery. Many oaks in the royal forests had been inadvertently damaged by foresters, who had cut from each tree a 6-inch square piece of bark so that the timber could be systematically branded and numbered. This had allowed the wood to become gravely rotted. Aware that Forsyth was using a method of curing defects in growing trees, the commissioners wrote to him inquiring if he could suggest "a complete remedy."

A good-natured, self-contained man, Forsyth had managed to reach high office through his abilities as a thoroughly capable professional gardener and not because of his formal educational attainments, of which he had none. By this stage in his career he appears to have become complacent, apparently thinking that he could survive indefinitely in the competitive world of public life without the need to choose his words carefully. It was quite possibly this, rather than any personal dishonesty, that drove him to respond as he did to the entreaty to save the nation's trees. There was perhaps also an incautiously gung-ho enthusiasm to help with the war effort. Thus he wrote back to the Treasury, claiming that he could restore oak trees "where nothing remained but the bark." Indeed, he boasted that such trees would be "rendered as fit for the Navy as though they never had been injured." Overeager to get his remedy accepted he needlessly drew attention to his lack of learning by informing them that the formula was "known only to myself, as it is not a secret drawn from books, or learned from men, but the effect of close application and repeated experiments."

Shortly afterward, in July 1789, an ad hoc committee of twelve members of both Houses of Parliament was appointed to ascertain the efficiency of Forsyth's remedy. The committee met one summer Saturday in the Royal Gardens, Kensington, where Forsyth showed them trees to which his "composition" had been applied in various experiments "for upwards of seven preceding years." Readily impressed, they reported to the Treasury: "From all we saw and heard, we have reason to believe that Mr. Forsyth's Composition is a discovery which may be highly beneficial both to individuals and the Public." The Treasury then recommended that the king should grant Forsyth £1,500 for the composition's formula on condition that he make the recipe widely known in writing to the public.

Accordingly, Forsyth placed a number of newspaper advertisements and produced a pamphlet, *Observations on the Diseases, Defects and Injuries of Fruit and Forest Trees* (1791), describing his methods in detail. The composition's recipe was as follows: "Take one bushel of fresh cow-dung, half a bushel of lime-rubbish of an old building plaster (that from the ceilings of rooms is preferable), half a bushel of wood-ashes, and a sixteenth part of a bushel of river sand: the three last articles are to be sifted fine before they are mixed, then worked well with a spade until the stuff is very smooth, like the fine plaster used for the ceilings of rooms." To apply the mixture, he instructed, "Lay on the plaster about one-eighth of an inch thick all over the part of the wood or bark that has been cut away," or, if it was to be applied in a liquid state, "it should be reduced to a consistency of a pretty thick paint, by mixing it up with a sufficient quantity of urine and soapsuds, and laid on with a painter's brush." No objections or criticisms were made when the pamphlet was published, nor when he subsequently developed it into his *Treatise on the Culture and Management of Fruit-Trees*, a hugely popular book which went through seven editions between 1802 and 1824.

A few years later wealthy landowner Thomas Andrew Knight was contemplating the science of grafting fruit trees at Elton, a 9,668-acre rural estate in Hertfordshire. In contrast to Forsyth, Knight believed that fruit trees had a predetermined finite life; he sought to establish this view by grafting branches from old trees onto young stock, and vice versa. The results apparently proved his theory since the graft responded according to the age of the donor tree, not of the recipient. Knight noted that when the graft came from an old donor it soon succumbed to canker, thereby meaning, he thought, that canker was solely a symptom of senility and therefore untreatable. In 1797 he published his findings in his *Treatise on the Culture of the Apple and Pear*. The book was subsequently scrutinized in the *Monthly Review*. This reputable journal regarded it with slight skepticism, and

AN OAK.

EXPLANATION OF PLATE XII.

THIS plate represents an old stunted oak, which was headed down about six years ago. At that time it was full of wounds and blemishes, now nearly healed.

a. The place where the tree was headed, afterwards covered with the Composition.

b, b, b. Three young shoots produced since heading ; there were several others, which were cut down as they advanced in growth : the two remaining side ones are also to be cut down, and only the middle one left, which will in time cover the wound *a,* and form a proper tree.

c, c, c. Remains of the old wounds, covered with the Composition, and now almost healed up.

Forsyth claimed his "Composition" could rejuvenate oak trees. (*Treatise on the Culture and Management of Fruit-Trees*)

hazarded the remark that Forsyth's work "gives strong reasons for suspecting that there may be some fallacy in Mr. Knight's hypothesis."

Though fortunate in his extensive landed inheritance, Knight was a temperamental man, quick to take offense and vindictive in furthering a grudge. To a man of his nature even such an innocuous remark could only be interpreted as an insult. Forsyth's position as a Pillar of the Establishment seemed to be secure. Indeed, as the royal gardener he was the gardening equivalent of the Poet Laureate! Knight, twenty years his junior and impatient for success, reckoned he could best make his own theory accepted by damaging his rival. He

164

visited Kensington Gardens incognito and artfully got Forsyth—who had never previously met him—to blurt out that he "did not place much confidence on any topical application to the wounds or diseases of vegetables." Knight then published an account of this alleged concession in the second edition to his own *Treatise on the Apple and Pear* (1801). A horticultural colleague of Forsyth's, Dr. James Anderson, the wealthy proprietor of *Recreations in Agriculture*, robustly sprang to the royal gardener's defense, claiming in that journal that Knight was guilty of willful misrepresentation.

Enraged, Knight issued a vitriolic pamphlet, *Some Doubts Relative to the Efficacy of Mr. Forsyth's Plaster in Filling Up Holes in Trees*, in which he made several wildly libelous allegations. He accused Anderson of being the ghost-writer for the uneducated Forsyth's book. Next he claimed the plastered fruit trees in Kensington Gardens were failing so dismally that Forsyth was buying fruit from Covent Garden market for the royal table. Furthermore, he said, Forsyth's claim to have invented the composition was fraudulent because the substance had been in use in rural areas for some fifty years before Forsyth received the parliamentary reward for his "discovery."

By this time Forsyth was at the pinnacle of his career, and on March 7, 1804, he met with other eminent personages in Hatchard's bookshop at 187 Piccadilly to found what was later to become the Horticultural Society. Nevertheless, some of the mud Knight was hurling was beginning to stick, and concessions were having to be made. In a new edition of his book Forsyth made the undignified admission that he might not necessarily have been the original inventor of the remedy, although the precise formula of his composition was his alone. The allegation that the composition was virtually useless was far more difficult to disprove, but Forsyth hoped to do so with the help of

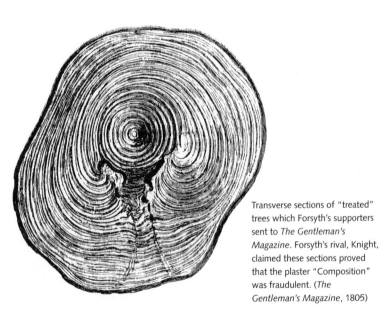

Transverse sections of "treated" trees which Forsyth's supporters sent to *The Gentleman's Magazine*. Forsyth's rival, Knight, claimed these sections proved that the plaster "Composition" was fraudulent. (*The Gentleman's Magazine*, 1805)

influential friends. The third edition of his book included a testimonial signed by six doctors who, having been shown the trees in Kensington, vouched for the truth of Forsyth's original 1789 statement. The prime mover behind these signatures was the eminent surgeon John Coakley Lettsom, a member of the Royal Society, who quickly found himself ridiculed for protecting Forsyth. The doctor's habit of signing his Latin prescriptions "I. Lettsom" soon gave rise to a popular rhyme:

> When any sick to me apply,
> I physics, bleeds and sweats 'em;
> If, after that, they choose to die,
> Why, verily! I. Lettsom.

When he saw this testimonial in print, Knight played his trump card. By means of an open letter in *The Gentleman's Magazine*, he wagered 200 guineas that Lettsom and the other signatories could not produce a transverse section of a tree that would substantiate Forsyth's claim of being able to restore hollow trees to perfect health. Then, on the verge of being unmasked as a fraud, Forsyth suddenly died at his official residence in Kensington on July 25, 1804. Left to salvage his reputation, his loyal supporters did eventually provide samples as illustrations for *The Gentleman's Magazine*. But when these were published Knight had no trouble in showing that the magic paste had done nothing to restore the oaks.

Forsyth's dishonor was by no means total, though it was Knight who went on to become head of the Horticultural Society. Yet in time doubts were raised about his own integrity. Why had he switched the argument to concentrate attention solely on the oak trees? In fact *The Gentleman's Magazine* subsequently found that Forsyth's plaster *had* been effective in renovating thousands of fruit and forest trees. But at the time of Knight's death in 1838 all the papers containing the evidence for his furious attack against Forsyth mysteriously disappeared, and have not been found since. As for Forsyth, it gradually became apparent that his claims to have done his own writing were probably truthful (when his son died in 1835 he left what had become the finest horticultural library in England). In other areas of tree care Forsyth had been a successful pioneer—he is believed to be one of the earliest modern growers to have noted that in certain species fruit buds grow far more quickly if the shoots are trained horizontally instead of vertically. Forsyth's supporters were forever loyal, and the Danish botanist Martin Vahl immortalized him by naming the cheerful yellow-flowering shrub *Forsythia* in his honor. It still serves as a tribute to a pioneer who was perhaps more of a fool than a knave.

 Forsyth's rockery is being restored as a listed building at the Chelsea Physic Garden, 66 Royal Hospital Road, London SW3 4HS. Website: www.cpgarden.demon.co.uk.

John Bennet Lawes's Superphosphate

The first ever advertisement for superphosphate appeared in *The Gardeners' Chronicle* on July 1, 1843. Inserted by John Bennet Lawes, it signified the beginning of the now extensive artificial fertilizer industry.

John Bennet Lawes was born on December 28, 1814, at Rothamsted Manor, a small agricultural estate on the outskirts of Harpenden in Hertfordshire, where his ancestors had lived since 1623. He was only eight when his father died and he inherited Rothamsted, but he continued his education at Eton before going on to study at Brasenose College, Oxford. The

Laweses were well-off but not wealthy, a circumstance which meant that, unlike many of the young squires who were his contemporaries at Oxford, he had no fortune to squander. As a student he developed an interest in chemistry and plant growth— an interest that was to alter the whole course of his life and the finances of his family.

J.B. Lawes. (IACR-Rothamsted)

Returning from the university to manage the Rothamsted estate in 1834, John converted the best room at the manor into a laboratory and commenced a series of chemical investigations. His early experiments involved the extraction of drugs from medicinal plants grown on his land. Acutely aware of the practical problems of agriculture, in 1837 he began investigating how farm crops reacted to various forms of manure. One of the manures Lawes tested was bone, an age-old, though slow-acting, form of phosphoric fertilizer. He was puzzled because bones, much valued for improving the yields of crops in some parts of the country, had little effect on his soil. In 1817 the Irishman James Murray and later the famous German chemist Baron Justus von Liebig had developed a theory which seemed to indicate that if bones were treated with sulphuric acid, the superphosphate by-product of that reaction could be a most effective manure. In 1841 Lawes put together and tested some small quantities for a field trial at the Rothamsted home farm. When it made his turnips grow more quickly, he became convinced that this new superphosphate could be of far greater value than had previously been realized.

He decided to manufacture the material on a larger scale, and set up his production line in a barn at Rothamsted. An "edge-runner"—a large heavy stone that rolled around a circular channel on a horizontal stone—was used to grind the bones brought from London. Since untreated bones did not grind well, Lawes usually purchased bones that had already been boiled or burned. Sometimes he used bone-dust, and even, reputedly, ivory powder. Sulphuric acid was carted to the farm in huge straw-covered jars, and when the ground bones were mixed with the acid, heat was produced, and the resulting mass set hard as it cooled. The edge-runner was then used again to grind the superphosphate. By this somewhat laborious process Lawes succeeded in making several hundred pounds of the world's first artificial fertilizer.

Realizing the enormous commercial potential of a compact soluble manure that could virtually replace farmyard manure, Lawes hastened to register his claim. On May 23, 1842, the Patent Office in London granted him a wide-ranging patent for the manufacture of fertilizers by three methods: sulphuric acid and bones; phosphoric acid and alkali; and silica and soda. He got there just in time. Later the same day other people unsuccessfully filed patents for various types of fertilizer mixes. Although Lawes was then in the process of getting married to Caroline Fontaine of Norford Hall, Norfolk, he refused to allow even that to slow the momentum of his business venture. Canceling their planned continental honeymoon, he took his blushing bride on a trip down the Thames—inspecting potential sites for his first factory.

By 1843 the superphosphate was being produced in vast quantities at the huge new manufacturing works at Deptford Creek, near Greenwich. The squire of Rothamsted was risking everything he had on this venture. Such was the scale of the

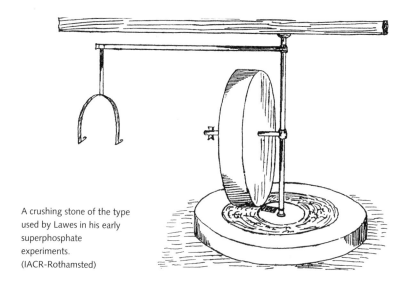

A crushing stone of the type used by Lawes in his early superphosphate experiments.
(IACR-Rothamsted)

output that for a while it seemed the available supplies of bone could not meet the demands of the factory. Then Lawes learned of the discovery of coprolite rock—the fossilized remains of prehistoric marine animals—in various parts of the world. This was a cheap and plentiful phosphate material that he successfully substituted for bone.

Another problem was that rival manufacturers were beginning to produce superphosphate. Lawes went to court to protect his rights under the patent. Control over two of the patented processes proved to be impossible to maintain, but the third, his sulphuric acid method, was upheld, and subsequently he was entitled to a royalty payment on all the superphosphate sold by his rivals—at a substantial 10 shillings per ton!

Despite his commercial activities Lawes never lost interest in agricultural experimentation. The profits of the now immensely successful business financed further trials at Rothamsted. Lawes was a true entrepreneur. Quick to see

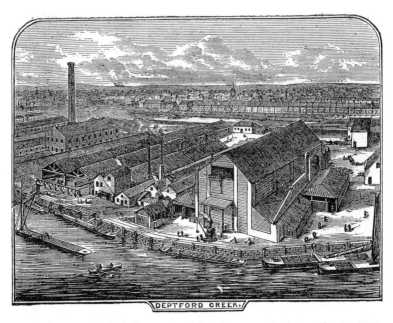

Lawes's factory at Deptford. He even canceled his honeymoon to discover the site. (IACR-Rothamsted)

what horticulturists and gardeners wanted, he provided sound practical and economic advice based on the results of the Rothamsted experiments. From 1843 the high quality of his results owed much to the diligent testing done by Dr. Joseph Gilbert, the supervising scientist there. Lawes's and Gilbert's work—particularly regarding the role of nitrogen in all plant growth—had a profound worldwide influence in placing the practice of horticulture on a sound scientific basis, and made the Rothamsted Experimental Station famous as one of the world's chief research centers. In 1872 Lawes sold his fertilizer business for £300,000, one-third of which he set aside to be administered by a trust for the continuance

The first advertisement for superphosphate in a gardening magazine. (IACR-Rothamsted)

of the experimental work. By the time of his death at Harpenden in 1900 Lawes was an unquestioned authority on most aspects of agricultural and horticultural science. For this he had been elected a Fellow of the Royal Society in 1854 and created a baronet in 1882.

The Lawes-pioneered manufacturing process spread far and wide across the world. The first American mixed fertilizer was patented in 1849 in Baltimore, which rapidly became an important center for the manufacture of superphosphates. Lawes's wonder-fertilizer was found to be ideal for making grass grow better in New Zealand where the soils were particularly deficient in sulphur. Superphosphate was the only manufactured fertilizer in England for nearly eighty years. In the process, Lawes's factory at Deptford did untold environmental damage: using nearly 200 tons of sulphuric acid a day, it polluted the Thames and ruined the health of both those who produced it and those who lived nearby. Nevertheless farmers and gardeners marveled for decades at the superphosphate's ability to hasten ripening and stimulate root development.

 For further information on John Bennet Lawes, contact the Integrated Approach to Crop Research (IACR), Rothamsted, Harpenden, Hertfordshire AL5 2JQ.

George Washington Carver's Peanut

Born into slavery and raised during the bitter years of the Reconstruction, George Washington Carver overcame unbelievable odds to become the leading agricultural chemist in America. The champion of the lowly peanut, he almost single-handedly revolutionized the Southern economy and freed the South from the crippling legacy of cotton.

George Washington Carver was born in about 1864 on a plantation near Diamond Grove, Missouri. Carver never knew his father; most likely he was a slave from a neighboring farm who was killed in an accident before his birth. When Carver was an infant, he and his mother were kidnapped by Confederate slave raiders. The owner of the plantation, Moses Carver, was able to recover George, but his mother was never seen again. After the Emancipation Proclamation, Moses Carver and his wife Susan raised George and his brother as foster children. George was an eager student and a lover of plants from an early age. Excused from fieldwork due to illness and frailty, the boy spent hours exploring the woods around the Carvers' farm, studying and collecting flora, and cultivating a small garden of his own.

At age twelve, barred from the local school by his color, he moved to Neosho, Missouri, to attend a school for blacks. After continuing his studies in Kansas, Carver in 1890 gained admission to Simpson College in Iowa, where he studied piano and painting. In 1891, at the urging of an art teacher who recognized his unique talent with plants, Carver enrolled at the Iowa State College of Agriculture and Mechanical Arts (today Iowa State University) to study agricultural science, the first black student ever to do so.

After receiving his B.S. degree in 1894, Carver was appointed to the college faculty and put in charge of the college greenhouse. Here he devoted himself full time to grafting and crossbreeding experiments. Louis Pammell, a distinguished botanist under whom Carver worked, described him as "a brilliant student, the best collector and the best scientific observer I have ever known." In 1896 Carver received an M.S. degree in botany and agriculture—an unprecedented accomplishment for a black citizen at the time. In the same year Booker T. Washington wrote to Carver inviting him to run the newly created Agricultural Experiment Station at the Tuskegee Institute in Alabama. Washington was a leading spokesman for black Americans who in 1895 had won national acclaim for a speech in which he urged the former slaves to acquire marketable skills and cultivate good relations with their white neighbors. Toward the first goal, Washington had in 1881 opened the Tuskegee Institute (today Tuskegee University) to train blacks to succeed in agriculture and the trades. In reply Carver wrote: "Of course it has always been the one great ideal of my life to be of the greatest good to the greatest number of my people possible, and to this end I have been preparing myself for these many years." Carver started at the Institute on October 8, 1896, with thirteen pupils under his tutelage.

In keeping with Tuskegee's philosophy, Carver set out to improve the lot of the millions of small-scale tenant farmers trapped in poverty in the South. As he saw it, one of his main obstacles was cotton. The aggressive cultivation of cotton during the nineteenth century had—in addition to driving the growth of slavery—exhausted the soil of the South, because cotton heavily taxes nutrients in the soil. When a field was no longer productive, planters would often move on to new fields, or burn down woodlots to create them. Carver described the phenomenon this way: "The virgin fertility of our soils and the vast amount of unskilled labor have been more of a curse than a blessing to agriculture. This exhaustive system for cultivation, the destruction of forest, the rapid and almost constant decomposition of organic matter, have made our agricultural problem one requiring more brains than that of the North, East or West."

At Tuskegee, Carver focused on the problem of soil conservation and developed ingenious crop-rotation methods to replenish the soil. He experimented with legumes: crops such as peas, beans, soybeans, peanuts, clovers, and vetches that enrich the soil with atmospheric nitrogen rather than depleting it. Carver discovered that by cultivating legumes and cotton in the same field in alternating years, farmers could maintain the productivity of the soil without the use of additional fertilizers. In addition to teaching students, Carver labored to educate common Southern farmers, both black and white, about the need to diversify their planting. He published articles and pamphlets and gave speeches and public demonstrations. Despite his efforts, and a declining demand for cotton, most farmers clung to the one-crop system, fearing that they could not sell other crops.

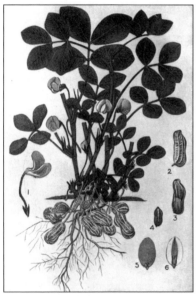

A botanical drawing of the peanut life cycle.

Around the turn of the century, the South was devastated by the boll weevil, a small beetle apparently evolved for no other purpose than to destroy cotton. Crop production was cut in half, and Carver's warnings found an audience. Farmers began planting cotton one year and the next year peanuts, a rugged crop of South American origin. The plan worked and soon large surpluses of peanuts piled up, bringing into focus the other half of the problem: how to find a market for the peanut and other crops that could complement cotton. It was in response to this challenge that Carver performed the work for which he is best known. In the following years Carver invented over 325 new commercial products based on the peanut, including cooking oil, cooking sauces, peanut flour, peanut butter, laundry soap, laxatives, cosmetics and lotions, shampoo, shaving cream, metal polish, axle grease, printer's ink, dyes, paints, and stains. From the sweet potato, the pecan, and other soil-enriching plants, he derived hundreds of other useful products. He did not seek to profit from or control his inventions, but offered them freely to any who could benefit. He held three patents only, all relating to a process he developed in 1927 for creating paints and stains from soybeans.

During his career at Tuskegee, Carver was not exclusively concerned with peanuts and soybeans. He also studied fungi, plant disease, and medicinal plants. He developed methods for analyzing soil and water, conducted pioneering research with fertilizers, and created several new strains of cotton, the most important of which bears his name. He also organized a highly effective and much-imitated community-outreach program, which included a mobile demonstration laboratory called the Jessup Wagon. By the time of his death in 1943 Carver was a legend. He received many honors, both during

George Washington Carver. (National Parks Service)

and after his life. In 1916 he was named a Fellow of the Royal Society of Arts of London. In 1939 he was given the Theodore Roosevelt Medal for his contributions to science. In 1951 a fifty-cent coin was created in his honor, and in the same year the George Washington Carver National Monument was established on the site of Moses Carver's farm, where he was born. Those who knew him described him as a gentle and spiritual man who cared little for material possessions. His epitaph reads: "He could have added fortune to fame, but caring for neither, he found happiness and honor in being helpful to the world."

 For more on George Washington Carver, see www.tusk.edu and www.nps.gov/gwca/expanded/index.htm.

GARDEN WRITINGS

Richard Bradley's *History of Succulent Plants*

In the early eighteenth century a growing stream of succulent plants from Africa, America, and the West Indies arrived in Holland, although few reached Britain. All that changed through the writings of the roguish Richard Bradley, a pioneering journalist and author—who was also quite a charlatan.

Richard Bradley's early life is shrouded in mystery. Born in 1688, he was brought up in the Cambridgeshire area, and from early childhood developed "a passion for gardening and planting." Of his student years nothing is known. Records show a Richard Bradley at Oxford University who was of the right age, but it was evidently not this Richard Bradley. Somehow, apparently without having received any verifiable formal education, by his early twenties he was making a living in London as a Grub Street hack. With his well-informed and comprehensive mind, and a natural journalist's flair for extracting information from the most unpromising individuals, he got by on his ability to write well. Having some knowledge of horticulture, he managed to get essays published on the subjects of moldiness in melons and the motion of sap. By 1710 he had become such a succulent enthusiast that he was collecting these rare and exotic plants; declaring his intention to produce an illustrated book about them, he printed and distributed a prospectus, *A Treatise of Succulent Plants*, in which he touted for investors.

He was elected a Fellow of the Royal Society in 1712 which brought him into friendly contact with many of the great naturalists of the era. Just as he had hoped, his admission to those privileged circles enabled him to obtain sponsorship from the Duchess of Beaufort and nurseryman Thomas Fairchild, which allowed him to acquire succulent plants from the Netherlands. Arriving there in 1714, he was introduced to the leading Dutch cultivators and other celebrities, and then purchased some two dozen of the sixty different African "aloes" available in Amsterdam. It must have been a thrilling experience for him to see so many novelty plants for the first time, and subsequently to introduce them successfully into England.

The first ever book in its field, *The History of Succulent Plants* was an ambitious publishing project. To overcome the problem of raising enough funds to publish the work outright, Bradley created a precedent by offering the work serially in five "decades" or segments. Ever the showman, he figured if he called the work a "History" it might sell better. When it appeared in 1717 it contained a series of fifty life-drawn illustrations. Bradley made these strange new plants accessible to his readers by providing them with picturesque names such as "Fig Marygold," and also included cultivation notes based on his own first-hand experience. Had public support been stronger he would have included all the succulents then in cultivation, but *The History of Succulent Plants* was never quite

Bradley's *History of Succulent Plants* introduced many of these beautiful plants to British readers.

the commercial success that he had anticipated (in 1717 he narrowly escaped being sent to a debtors' prison).

By 1714 Bradley had a wife—whose name he never revealed—with whom he was living in a palatial residence in Kensington, where he tended his private collection of rare succulents. A shifty character, who always seemed to have some sort of money-making scheme going on, Bradley made ends meet by taking other jobs on the side. One such scam involved him in overseeing the planting of the Duke of Chandos's garden at Canons—the plants for which he incompetently failed to deliver. Then, in 1720, disaster struck. Already financially stretched, he was ruined by the South Sea Bubble disaster.

Ever resourceful, at the age of only twenty-six he succeeded in getting himself elected Professor of Botany at Cambridge.

Sturt sc.

The book had pictures of fifty striking exotics. Many had never been seen in England before.

This was, however, a pyrrhic victory because although the status enhanced the salability of his writings the position was unpaid. Capitalizing on his title he became one of the earliest celebrity gardeners to have his name on books ghosted by others. These included a number of translations, and covered fairly obscure subjects such as economic theory. Another milestone was his *Monthly Register of Experiments* which survived only a year, but is believed to have been the first ever horticultural journal available on subscription. Bradley was a talented and proficient plant illustrator, and in devising a means of producing new geometrical shapes for landscape gardens, he effectively became the inventor of the kaleidoscope.

Little is known about the final years of Bradley's life. He was living at Stratford in Essex in 1729, and in 1730 he remarried, this time to a woman of "considerable fortune" called Mary. Her wealth quickly evaporated. They had one child before he became seriously ill, and he died, poverty-stricken, in November 1732 in lodgings on Charterhouse Lane. His young wife had no choice but to write begging letters and sell their belongings in order to pay their debts. It was a tragic end for a writer who had been so well known and generally highly regarded during his lifetime.

After his death, Bradley's reputation fell into disgrace, largely because of the aspersions cast on his moral character by another Cambridge professor, John Martyn. Martyn bore a bitter grudge against Bradley, partly because he envied his literary success, but partly because he thought Bradley had cheated him out of the first Cambridge chair of botany. He alleged that Bradley had secured the professorship in 1724 under false pretenses; that he had reneged on a promise he had solemnly given to build Cambridge a botanical garden out of his own pocket; that he was unable to write any Greek; and that he never even bothered to give any lectures. Nevertheless, despite these allegations, Bradley deserves a place in gardening history. *The History of Succulent Plants* was the only book on the subject until 1794, and laid the foundations for the study of succulents both in England and abroad. Interestingly, of the fifty succulents that Bradley introduced to the British public, all are still in cultivation.

 For further information on succulent plants, contact www.succulent-plant.com.

Carl Linnaeus and His Plant Classification

Gardeners have good reason to be grateful to Carl Linnaeus. Before his time the long descriptive Latin titles appended to plants, the profusion of synonyms, and the confusion of species and varieties were rapidly producing chaos. Angry customers were writing indignant letters to nurserymen when they found they had purchased expensive plants that they already possessed under another name.

Born in 1707 at Rashult, a village near Lake Mocklen in southern Sweden, Carl Linnaeus was the eldest son of a country clergyman. His youth was spent amid an idyllic profusion of flowers, some native and wild in the nearby marshland and meadows, others exotic and cultivated in his father's well-stocked ecclesiastical garden. When he was five years old Carl would work with his father in the garden, constantly asking him the names of plants. Yet he so frequently forgot them that his exasperated father refused to tell him any more! Linnaeus set his heart on remembering the names, and thereby began his

Carolus Linnaeus. (Reproduced by permission of the Royal Botanic Gardens, Kew)

lifelong enthusiasm with the naming of organisms. His interests were so focused on natural history that his progress in the studies necessary for admission to the church, for which he was intended, faltered and in 1726 his father was advised to apprentice him to a tailor or a shoemaker. Linnaeus was narrowly saved from that mediocre fate by a local physician in the town, who hazarded a belief that Carl would yet distinguish himself in medicine and natural history.

In 1729 the penniless Linnaeus was studying medicine at Uppsala University but that autumn everything unexpectedly changed. The Professor of Botany, Olaf Rudbeck the younger, learned that the desperate student had applied for a job as keeper of the university's neglected botanical garden. Originally laid out in 1653 by Rudbeck the elder, the professor's father, it was the oldest botanical garden in Sweden. Not only did he get the job, but Linnaeus was invited to lecture on the plants in the garden and thus began his long career as a university teacher. During this time, he was strongly influenced by Sébastien Vaillant's book *Sermo de Structura Florum* (1717), which encouraged him to examine the stamens and pistils of flowers. Becoming convinced of the paramount importance of these organs, Linnaeus formed the idea of basing upon them a system of classification. Thus, by the age of twenty-four, he had laid the foundations of all his later work. Significant personal qualities had also developed, notably his quite remarkable mental stamina and tenacity, a strong visual memory, and an exceptionally methodical mind. Furthermore, arrangement had become a passion—he simply delighted in devising categories.

The first of Linnaeus's two great contributions to botany was a sexual system of plant classification based on the generative organs. This he propounded in his *Systema Naturae* (1736),

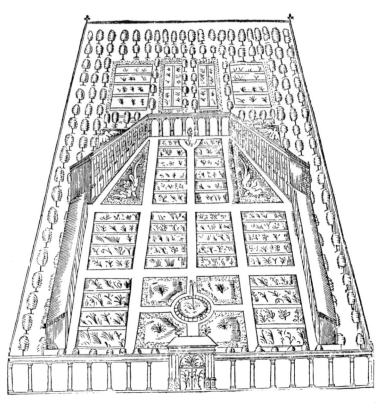

The Botanical Garden in 1679, long before Linnaeus began working there. (Linnean Society of London)

introducing it as a system for classifying plants into groups by categories such as the number of stamens in the flowers. This provided a much-needed framework for easy identification. The second, and more important, element of the "Linnean system" appeared in his *Species Plantarum* (1753), in which he established an entirely new binomial system of nomenclature. His idea was that every living creature should be simply and concisely distinguished by *two* internationally recognized Latin names—a generic name followed by a specific name. For example, in the Latin name for the daisy, *Bellis perennis*, *Bellis* is the name of the genus to which the plant belongs (written with an initial capital letter), while *perennis* distinguishes the species from others of the same genus. The author who first described a particular species would be indicated after the name (which could be abbreviated). Thus *Bellis perennis L.* showed that the author was Linnaeus.

Many of Linnaeus's contemporaries were fairly hostile at first to this so-called "sexual system" of classification. Prior to this there had been a "natural" system of classification based on the belief that the Creation itself must have been orderly and systematic. The Linnean formula meant changing these quaint well-established generic names. Another factor was professional jealousy since Linnaeus eventually had many thousands of plants associated with his name—a figure that exceeded even the numbers attributed to Eve (who is supposed to have undertaken that part of the work when Adam named the animals).

Making a meteoric rise to the head of his profession, Linnaeus succeeded the younger Rudbeck as Professor of Botany at the University of Uppsala in 1742. He was just thirty-five years old. His lectures drew scholars from all over the world; the normal number of students at Uppsala had been five hundred, but while he occupied the chair of botany there this number trebled. He spoke no contemporary language except his own, and deliberately wrote, conversed, and lectured to many of his students in Latin—the *lingua franca* of the learned. The atmosphere at Uppsala was alive with a sense of momentous change. He imbued the students with his own inquisitiveness, and having trained them to observe closely and accurately he dispatched them to various parts of the globe.

Twenty-three of Linnaeus's students themselves became professors and thus spread his methods widely. Two of the Linnean "apostles," Daniel Solander and Andreas Sparrman, even sailed around the world with Captain Cook.

In addition to his botanical attainments, Linnaeus was a physician far in advance of his time and he also did much valuable work in zoology, entomology, and mineralogy. He was also the first European to discover a method of making cultured pearls. Ennobled in 1757 in Sweden, where he was a national hero, he became Carl von Linné. After his death in 1778 his extensive library and personal herbarium were sold by his widow to the Linnean Society of London. The Linnean system remains the accepted basis of botanical nomenclature, and so to this day his influence is apparent on plant labels in garden centers and in millions of gardens around the world.

 For more details on Carl Linnaeus's plant classifications, contact The Linnean Society of London, Burlington House, London W1; www.linnean.org.uk.

Jane Loudon's *Gardening for Ladies*

For centuries, insofar as books had been written for female gardeners, their sole purpose had been to show how ladies might oversee the tending of flower and herb gardens. When it appeared in 1840 *Gardening for Ladies* broke new ground by being the first ever practical hands-on guide for green-thumbed gentlewomen. Amazingly, the book—which did for the outdoor activities of inexperienced mistresses of Victorian households what Mrs. Beeton did for indoor economy—was created by a pioneering female science-fiction writer.

Mrs. LOUDON'S

GARDENING

FOR

LADIES.

The frontispiece of Jane Loudon's revolutionary *Gardening for Ladies*, published in 1840.

Jane Webb was born in 1807 at Ritwell House, near Birmingham, the ancestral home of her well-to-do parents. When her mother died Jane became mistress of the large family house with its substantial flower garden and some 30 acres of grounds. Her father, in consequence of overspeculation, became financially embarrassed. When he died in 1824, Jane decided she would earn her living as a writer and she soon published a book of prose and verse which was chiefly about a year she had spent wandering around the continent of Europe. Possessed of a lively mind and a well-balanced temperament, she took a keen interest in the practical application of technology—an unusual pursuit for a woman at a time when ladies were not expected to have original ideas. Her next venture was *The Mummy*, a futuristic novel set in the twenty-second century, in a world in which she foresaw air travel, wireless, telephones, air conditioning, and mechanical milking machines. In 1828 the book caught the attention of John Claudius Loudon, editor of *The Gardener's Magazine*, who was so fascinated by the notion of a steam-driven plow that he wrote a favorable review of the book under the heading "Hints for Improvements." Two years later Jane met Loudon at a party. They fell in love, and within months they were married.

Born at Cambuslang in Scotland in 1783, John Loudon had initially made a name for himself as the founder of the first agricultural college—a profitable business which he then sold so that he could turn his attention to writing. His health was poor, and he was plagued by chronic rheumatism which resulted in his right arm being amputated when he was only twenty-five. But his writing career blossomed, and he managed to escape severe financial trouble. His *Encyclopaedia of Gardening* (1822), a three-year intensive project, which was the first truly comprehensive manual of gardening, became a massive success, going into several reprints. With the proceeds from this he designed and built himself an impressive villa in Porchester Place, Notting Hill, and it was here, in 1826, that he established offices for his new publication, *The Gardener's Magazine*. This too was phenomenally successful. It was aimed at a new audience, concentrating on the smaller domestic gardens of the upper-middle classes rather than the gentry's country estates. In an effort to raise the intellect and character of its readers, it gave particular attention to the application of scientific and mechanical developments to gardening. With crusading zeal Loudon used it to denigrate the picturesque garden style in favor of what he termed the "gardenesque" patterns which celebrated the display of individual plants to best advantage and embraced orderly diversity.

Irrespective of the fact that John was twenty-four years older than Jane, they were eminently complementary: her calm, cheerful robustness proved the perfect counterpoint to his domineering dour demeanor. When he traveled about the country visiting gardens, she would invariably accompany him, whatever the weather, often bringing along their young daughter Agnes. Jane became Loudon's closest confidante and his literary assistant; she even provided some reviews and illustrations for his magazine.

Their status as the first family of gardening throughout the 1830s did nothing to lessen the gathering financial storm-clouds. The introduction of Joseph Paxton's *Horticultural Register* provided the first serious rival to their magazine, which had previously given them a comfortable income. In 1833 John Loudon's great success in publishing an encyclopedia of rural architecture catastrophically tempted him to produce an encyclopedia of trees on his own account. The *Arboretum* grew disastrously from one volume to eight, but Loudon pushed on with it in the hope of recouping his growing losses. He personally supervised his seven artists (who eventually produced some three thousand illustrations), and collated all the information gathered by himself and over eight hundred correspondents. Finally published in 1838, the *Arboretum* was a total flop. Subscriptions failed to reach Loudon's ever-sanguine expectations, and the contributors now demanded payment, meaning he found himself with a debt of some £10,000 which he could not discharge. His family rallied round. Two of his sisters learned the art of wood-engraving to spare him the expense of employing an engraver for his illustrations. To help sustain him, Jane threw herself into resuming her career as an author on her own account. This expedient move was the start of her own career as a famous horticultural writer.

The result was *Gardening for Ladies*, a 96-page book which for the first time showed women how to perform strenuous manual tasks in the garden. For example, Jane explained how to choose the right sort of willow-handled lightweight spade, which would enable them to shovel like manual laborers, and how to protect their hands with leather gauntlets (which she had designed). She also unveiled the mysteries of manure. All this was virgin ground for dainty-natured upper-class women whose soft white hands had never yet come into contact with the soil. In other chapters she dealt with similarly earthy matters such as pruning, sowing seeds, and propagation.

It made no difference that the book had come about by force of circumstance. The timing was perfect. Eager to imitate the grand ladies whose families owned historic gardens, there were vast numbers of middle-class women living in prim Victorian villas who wished to play a more active part in the management of their own new flower gardens. To help them in this laudable

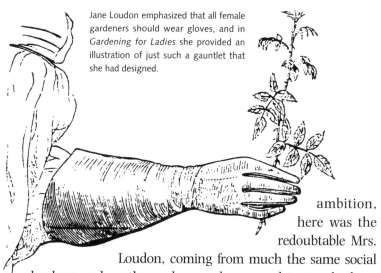

Jane Loudon emphasized that all female gardeners should wear gloves, and in *Gardening for Ladies* she provided an illustration of just such a gauntlet that she had designed.

ambition, here was the redoubtable Mrs. Loudon, coming from much the same social background as themselves and prepared to teach them everything they needed to know about gardening. She struck all the right notes, dutifully dedicating the work to her eminent horticulturist husband "to whom the author owes all the knowledge of the subject she possesses." With reassuring modesty she claimed that she had only really mastered her subject within the last few years.

Gardening for Ladies, with its pretty green leather cover engraved with gilt decoration, scored an immediate hit on publication. Over a thousand copies were sold on the first day, and it subsequently went into many editions. The influence of *Gardening for Ladies* spread to America where Andrew Jackson Downing wrote a foreword to a special edition—which left out the chapter on vegetables for being too proletarian! Despite this success, the financial plight of the celebrated Loudons was worsening. Creditors threatened to seize their Notting Hill home. Loudon had taxed his health far beyond his capacity and, distraught with financial anxiety, he

Nos. 3 and 5 Porchester Place, Jane and John Loudon's home and office. The house was designed and built by John Claudius Loudon.

collapsed and died while dictating his last book to Jane, by candlelight, at midnight on December 13, 1843.

To capitalize on her success as a garden writer, Jane had launched and edited the *Ladies Magazine of Gardening*, a journal which unfortunately survived only eleven months. Undeterred, she developed a successful *Ladies Flower-Garden* series, characteristically giving cheerful, unaffected, matter-of-fact instructions and doing the illustrations herself. These became standard works of reference and attained a large circulation. In total she wrote nineteen books on horticulture and natural history, many of which were produced by lithography, the new process that provided the first comparatively cheap method of color printing. Her books appeared on all Victorian bookshelves. Her income from writing, a Civil List pension, and help from friends enabled her to remain in the Porchester Place house until her death in 1858. It was largely through her writing and encouragement that gardening became an essential pastime for women, and has remained so ever since.

 Useful website which mentions Jane Loudon: www.gardenvisit.com.

James Shirley Hibberd's *Rustic Adornments*

Following the demise of John Loudon, for at least part of the 1860s the advocate of progressive gardening was James Shirley Hibberd. His opinions were influential and his name a household word. Yet the social world of gardening was such that his championing of the lower middle classes in the suburbs soon led to his being patronized and overshadowed by more fashionable writers, and he was later unfairly airbrushed out of the history books.

Born in Stepney, East London, in 1825, James Shirley Hibberd was the son of a retired sea captain. His mother might perhaps have been a member of the East End's property owning Shirley family, which may explain why he chose to abandon his first name, preferring to be called Shirley. James was only fourteen when his father died. As a result, he was obliged to take up a trade instead of, as had been intended, entering the medical profession. Apprenticed to a Stepney bookseller he began to write articles, although he also became a self-taught chemist. By 1850 he was editor of the official newspaper of the Vegetarian Society. Hibberd was a natural communicator and he expressed his talent widely, writing various pamphlets and also giving public lectures on subjects as diverse as popular science, history, and aesthetics.

James Shirley Hibberd. (*Journal of Horticulture and Cottage Gardener*, November 1890)

Hibberd did not originally set out to be a horticultural writer. That came about partly because of personal circumstance. He and his wife Sarah, who was a semi-invalid with a weak heart, had no children. If there was an emptiness in their lives she compensated for it by making their small house in Pentonville, North London, a menagerie, with pets of every description, while James energetically worked off his professional literary frustrations by exerting himself in the garden. When his literary career did successfully take off in 1855, it was largely due to *The Town Garden*, a book published that year and largely based on his experiences of starting and cultivating the Pentonville patch. The enthusiasm with which it was greeted revealed that he had tapped into a broad and previously overlooked vein of eager amateur gardeners. These were the lower middle classes, who lived in the terraced and semi-detached houses of subtopia, between the city offices and the suburbs. This was the world of frugal, upwardly mobile clerks who, though decent and ambitious, were lampooned by the working classes for their aloofness, and sneered at by the suburban middle classes for their clumsy lack of social grace.

The same audience welcomed Hibberd's *Rustic Adornments for Homes of Good Taste* when it was published in 1856. This 500-page book suggested an entire range of creative and interesting pursuits outside, and indeed inside, the "Home of Taste," and covered such items as aquariums, plant cases, floral ornaments, rockeries, and ferneries. Other books had been written on such subjects, but Hibberd's succeeded by gathering all the ideas into one volume. Even if some of the accessories he suggested—aviaries, beehives, and ornate summerhouses— were absurdly grand for diminutive urban gardens, they were, he said, affordable for people of narrow means. Thus, even the working class could experience, and be improved by, the pleasure and dignity of garden culture. He wrote:

In the Rustic Adornment of the home, it does not require a princely fortune to set up a vase of flowers, or an Aquarium, or a stand of bees that sing to their master all day long, and entrap every spare moment of leisure he is able to shepherd them. He who sets out his garden in accordance with correct principles of taste, will find in it as much amusement, and as genuine a solace from the cark and care of life, as if it were a domain of thousands of acres—perhaps more so. For it is his own work, it represents his own idea, it is a part of himself, and hence redolent of heart-ease.

The success of *Rustic Adornments* secured Hibberd's career as a serious garden writer, and in 1858 he began editing *Floral World*, a monthly newspaper aimed at owners of small gardens and people with modest incomes. Other popularist books then flowed from his pen, on topics ranging from greenhouses, ferns, and seaweed to wildflowers, roses, and ivies.

Quite unfazed by these literary triumphs, he remained determined to build on his strength as an individual writer who communicated practical advice derived from close personal experience. He also moved to a house in Stoke Newington, another subtopian London borough, where he set out to tackle the problems of urban and suburban horticulture. His new garden was an oblong strip, 282 feet long and 36 feet wide, and he ran a plan of it in the magazine's first volume of 1858. Issue by issue readers were informed with anecdotes and stories of his progress in transforming the unpromising site into a Shangri-La. Hibberd received so many letters from people wishing to visit this horticultural soap opera that he had to put a note in the magazine declaring it was an experimental plot not open to the public. It signifies something of his strength of purpose

Hibberd's celebrated house and garden in Stoke Newington. (Hackney Archives)

that when he became the editor of *The Gardeners' Magazine* in 1861—confusingly it was quite different from Loudon's lapsed *Gardener's Magazine*—he chose not to lead a more glamorous lifestyle. Instead, at his own expense, he started experimental gardens in other unfashionable suburbs to run trials of fruit trees and vegetables, notably potatoes, and engaged in all sorts of comprehensive flower and vegetable trials to provide practical and useful data for his readers.

Already this eminently reasonable and respectable writer had been depicted by contemporary horticulturists as being the wildest of eccentrics. Hibberd was a teetotaler and a vegetarian at a time when it was regarded as odd to refuse alcohol, and excessively Bohemian to be vegetarian. An animal lover, he wanted kindness shown to cattle in slaughterhouses. In garden warfare he was a pacifist who refused even to discourage birds or grubs, irrespective of how much damage they might do. There were other ideas by which—such was his reckless

EMBELLISHMENTS OF THE GARDEN.

THIS quiet garden's humble bound,
  This homely roof, this rustic fane,
With playful tendrils twining round,
  And woodbines peeping at the pane.

That tranquil, tender, sky of blue,
  Where clouds of golden radiance skim,
Those ranging trees of various hue,—
  These were the sights that solaced him.
                    TAYLOR'S Visit to Cowper's Arbour.

O embellish a garden well, needs a discrim-
inating and in some cases, a severe taste.
Whatever errors may be committed in the
laying out, the planting, and the disposition
of colours, will more readily escape the eye
or meet with forgiveness from the critic, than
the injudicious adoption of any kind of special embellishment.
Yet if the leading principles of gardening taste are kept in

In *Rustic Adornments* Hibberd insisted that even suburban gardeners of modest means could aspire to the finer things of life.

fearlessness—he made himself a laughingstock. These included his advocacy of such gardening innovations as holding fruit branches down with weights to avoid the need for pruning, and growing potatoes on tiles.

On a personal level, after the death of his first wife, Hibberd found happiness again in 1884 when he married his cook, Ellen Mantle. Aged twenty-eight, she was thirty-one years his junior. Hibberd died suddenly in 1890 and so, almost as quickly, did much of his good standing. Sadly, his memory was overshadowed by the formidable Gertrude Jekyll.

Useful website for information on John Shirley Hibberd:
www.hackney.gov.uk.

William Robinson's *English Flower Garden*

The most influential gardening book ever produced on English gardens was actually written by an Irishman. William Robinson was born in County Down on July 15, 1838. A Protestant from a very humble background, virtually nothing is known of his earliest years except that he had an elementary education at a parish school. After a brief apprenticeship on the estate of the Marquess of Waterford at Curraghmore, he rose to the position of foreman on a gentleman's estate at Ballykilcavan, where he had control over several large conservatories and hothouses. William's pugnacious fiery temper became all too apparent when he stormed out after a quarrel—stopping only to open all the doors and windows of most of the hothouses on a frosty night, thus killing his employer's valuable collection of exotics.

At the age of twenty-three Robinson moved to London and joined the Royal Botanical Society's garden in Regent's Park, where he was appointed foreman in charge of the herbaceous plants. His travels about the country in order to obtain stock for that garden made him familiar with the beauty of wildflowers commonly found in English cottage gardens. In 1867 he was sent as a correspondent by *The Times* to the Paris Exposition, where he also represented the firm of Veitch. In a tour through France he was quick to appreciate the practical French methods of cultivating fruit and vegetables, and on his return to London in 1868 he issued a publication called *Gleanings from French Gardens*. Next he produced his first important book, *The Parks, Gardens, and Promenades of Paris*. During his French tour he had also explored the Alps, and in order to spread the knowledge of a branch of gardening then little understood, he published in 1870 *Alpine Flowers for English Gardens*. With the money made from these books he set off on a visit to North America, where he studied the wonderful flora of the Rocky Mountains and California, and thereafter never ceased to advocate the use of North American trees and shrubs in English gardens.

His articles in *The Times* made him the outstanding gardening authority in Britain and in 1871 Robinson founded a magazine, *The Garden*. As its editor he gathered around him a host of enthusiastic practical amateurs, including Gertrude Jekyll, and devoted the magazine to promoting "pure horticulture of the natural, or English school, free from rigid formalities and meretricious ornaments." Robinson was a skillful writer, and a combative one. His interest was in horticulture, and its subordination to architectural values he saw as the degradation of gardening; he also became a pioneer of the rock garden.

His book *The Wild Garden* (1870) promoted the planting of exotic plants that would naturalize themselves in English gardens, and could be maintained with little effort.

The Subtropical Garden (1871) showed middle class gardeners how to achieve the effects of foliage gardening with hardy plants. *Hardy Flowers* (1871) helped to publicize the cause with which Robinson's name was to be mainly associated—the use of hardy plants in the garden. In 1879 he created a further opportunity to make outspoken criticisms of Victorian fussiness and formality when he established the popular magazine *Gardening Illustrated.*

The English Flower Garden was published in 1883. In its first edition it was to a large extent an anthology of the writings of prominent gardeners of the day. Written with a polemical emphasis, the book attacked the views of the champions of formal gardening. In it Robinson advanced his crusade against the privileged order. It was the custom of owners of country houses to allow their gardens to lie bare and dismal for nine months of the year, in order that they might enjoy a pyrotechnic floral display during the other three. Robinson showed that by using plants from the temperate regions of the world it was possible to have a garden that was bright for eight months of the year and charged with interest in the other four. With *The English Flower Garden* Robinson instigated a revolution in gardening. In his own inimitable style he directed the thoughts of Victorian and Edwardian gardeners toward the more informal and natural trends of garden making.

Enriched by the sales of this book and the success of his magazines, in 1884 Robinson bought Gravetye Manor, an Elizabethan mansion in Sussex with a 200-acre estate. Here he spent the next fifty years (the last thirty wheelchair-bound) working on the garden, writing, and advising others on their gardens. With the assistance of Gertrude Jekyll, he fought manfully in his numerous books and periodicals for the wild or wooded garden—an innovation of his own—and for the

THE

ENGLISH FLOWER GARDEN

STYLE, POSITION, AND ARRANGEMENT;

FOLLOWED BY

A DESCRIPTION OF ALL THE BEST PLANTS FOR IT;

THEIR CULTURE AND ARRANGEMENT.

FORMING VOL. I. OF THE 'GARDEN CYCLOPÆDIA.'

BY W. ROBINSON,

FOUNDER AND DIRECTING EDITOR OF "THE GARDEN," "FARM AND HOME," AND
"GARDENING ILLUSTRATED;" GARDEN EDITOR OF "THE FIELD;"
AUTHOR OF "THE PARKS AND GARDENS OF PARIS," "THE SUBTROPICAL
GARDEN," "ALPINE FLOWERS FOR ENGLISH GARDENS,' "THE WILD
GARDEN," "HARDY FLOWERS," ETC.

ILLUSTRATED WITH MANY ENGRAVINGS.

SECOND EDITION.

LONDON:
JOHN MURRAY, ALBEMARLE STREET.
1889.

The title page of William Robinson's phenomenally successful *The English Flower Garden*.

205

mixed herbaceous border. Although partly paralyzed by a stroke in 1910 he maintained his interest in gardens until his death in 1935 and regularly toured his estate in a chauffeur-driven caterpillar tractor. Robinson was planning a new orchard at the age of ninety-five, and was still an avid reader of English and French literature. He was irascible to the end, and books he considered to be "unworthy of good paper and print, and well-bound" he would throw onto his great log fire.

From 1883 onward, for the remaining years of his life, preparation of new editions of *The English Flower Garden* kept Robinson continuously busy. Changing economic and social circumstances meant his teaching subsequently became more appropriate than ever before. Less money was available for garden maintenance. Laborers were more difficult to find. Also, as thoroughly trained gardeners became scarcer, the elaborate bedding schemes of the Victorians grew even further out of reach than they had been when Robinson began to present his famous doctrines. *The English Flower Garden* became one of the classics of gardening literature. It guided generations of gardeners and gave untold pleasure to thousands of flower lovers. Effectively the inventor of the herbaceous border, the wild garden, and the alpine rock garden, Robinson's impact on twentieth-century gardening was immense. In fact, in 1933 Prime Minister Ramsay MacDonald seriously considered offering a knighthood to the "Father of the English Flower Garden."

 The site of the Royal Botanic Society's gardens, where Robinson trained for a while, can be seen at the Inner Circle, Regents Park, London NW1. His influential book, *The English Flower Garden*, can be purchased as a reprint, and his Gravetye Manor is now run as a country hotel at Vowels Lane, East Grinstead, East Sussex RH19 4LJ.

🌿 Robert B. Thomas and *The Old Farmer's Almanac*

American farmers and gardeners have been referring to *The Old Farmer's Almanac* for nearly as long as they have been Americans. The almanac was founded in 1792 as *The Farmer's Almanack, Calculated on a New and Improved Plan* by Robert B. Thomas, a schoolteacher, bookseller, and astronomer living near Boston. Since then the almanac has appeared every September, delivering its unique blend of astronomical and tidal data, farming and gardening advice, humor, folklore, remedies, recipes, and of course the daily weather predictions, prepared a full year in advance. Today *The Old Farmer's Almanac*, based in the tiny town of Dublin, New Hampshire, is the nation's oldest continuously published periodical.

The Old Farmer's Almanac is the longest-surviving example of what was once a crowded genre. By definition an almanac is a calendar for the coming year, together with weather forecasts, astronomical information, tide tables, farming advice, and the like. (The origin of the word is uncertain, but it seems to come from the Arabic word *al manakh*, meaning

Robert B. Thomas. (Yankee Publishing Inc., Dublin, New Hampshire)

"reckoning.") For the mariners and farmers who first landed on the coast of North America, the almanac was an essential tool. The continent's first printing press arrived in Cambridge, Massachusetts, in 1638, and the second publication it turned out was an almanac. During the eighteenth century, dozens of publishers put out competing almanacs, among them Benjamin Franklin's bestselling *Poor Richard's Almanack*, which appeared from 1732 to 1757. American farmers relied on such books for everything from planting tips to medical advice to entertainment.

Despite stiff competition, Thomas's *Farmer's Almanac* was a hit from the start; its circulation jumped from three thousand to nine thousand after a single year in print. Legend has it that Thomas's success was due in part to a secret weather-prediction formula, based on complex natural cycles, that allowed him to predict the weather accurately a year in advance. Since his format was not novel, it's generally believed that Thomas's astronomical and weather predictions were more accurate, his advice more useful, and his features more entertaining than other almanacs. When Thomas died in 1846 at the age of eighty (supposedly reading galleys for the 1847 edition), he had served as editor for fifty-four years and his almanac had become America's leading publication.

Subsequent editors did little to alter Thomas's creation. The second editor, John H. Jenks, changed the title to *The Old Farmer's Almanac* in 1848, and in 1851 introduced the classic cover engraving of Benjamin Franklin and Robert B. Thomas together with the four seasons, a design that has been used ever since. In 1936, however, the venerable Boston publishing company Little, Brown & Co. took over the almanac and named one of their own, Roger Scaife, as the

[Nº. I.]

THE

FARMER's ALMANAC,

CALCULATED ON A NEW AND IMPROVED PLAN,

FOR THE YEAR OF OUR LORD
1793 :

*Being the firſt after Leap Year, and ſeventeenth of the
Independence of* America.

Fitted to the town of BOSTON, but will ſerve for any of
the adjoining States.

Containing, beſides the large number of ASTRO-
NOMICAL CALCULATIONS and FARMER'S CA-
LENDAR for every month in the year, as great a vari-
ety as are to be found in any other Almanac,
Of NEW, USEFUL, *and* ENTERTAINING MATTER.

BY ROBERT B. THOMAS.

" While the bright radient ſun in centre glows,
The earth, in annual motion round it goes ;
At the ſame time on its own axis reels,
And gives us change of ſeaſons as it wheels."

Publiſhed according to Act of Congreſs.

PRINTED AT THE Apollo Preſs, IN BOSTON,
BY BELKNAP AND HALL,
Sold at their Office, State Street ; alſo, by the *Author*
and *M. Smith, Sterling.*
[*Sixpence ſingle,* 4*s. per dozen,* 40*s. per groce.*]

The title page of the first issue of the *Almanac.* (Yankee Publishing Inc., Dublin,
New Hampshire)

tenth editor of the publication. A member of the Boston literary establishment, Scaife secured contributions from the likes of Robert Frost, but he committed an astonishing blunder: In the 1938 edition he dropped the weather forecasts, leaving in their place only timid averages of temperature and precipitation. Readers were outraged and the forecasts were quickly reinstated, but the damage was done. During Scaife's tenure circulation dropped precipitously, and by 1939 Little, Brown & Co. was ready to sell. The man who bought the ailing almanac (he recalls being handed two double martinis along with the contract and a pen) was Robb Sagendorph, who had moved to Dublin, New Hampshire, four years earlier to start *Yankee Magazine.* Believing that the almanac's appeal lay in its tradition and a growing American appetite for nostalgia, he returned the magazine to its original format and tone. The steady growth of the almanac after World War II would prove him right. In the early 1990s, circulation passed the four million mark.

The modern *Old Farmer's Almanac* still closely resembles the little book that Robert B. Thomas first printed, although it has fattened to around three hundred pages. The right-hand pages of the calendar are crammed with holidays, holy days, historic dates, feasts, folklore, and tidal heights, each in a signature font. Weather for the month is summarized by the "weather prediction rhymes," or doggerels, that snake down the right side. The left-hand page of the calendar tracks the rising and setting of the sun and moon, the comings and goings of the planets, and the ebb and flow of the tides. According to legend, the young defense attorney Abraham Lincoln used this page in the 1857 edition to clear a client of a murder charge. It was argued that the alleged murderer could have identified the victim by the light of the moon;

Lincoln produced a copy of *The Old Farmer's Almanac* and easily satisfied the jury that the moon had been "riding low" that night.

Over the years, as the number of Americans involved in farming has declined dramatically, the almanac has addressed itself more and more to gardeners. The large gardening sections of the modern almanac, less playful than some of the features on topics such as goat breeding and how to tell when someone is lying, convey an enormous amount of useful information to the reader, much as the almanacs of the past did. Vegetable- and herb-growing charts teach the novice how to plant and grow a complete kitchen garden. The frost calendar charts the first and last frosts for dozens of U.S. cities. Reference tables list everything from the pH preferences of dozens of plants, trees, and shrubs to the flowers that best attract butterflies. Outdoor planting tables provide the recommended planting dates for thirty-six common vegetables and plants, as well as the most favorable moon for planting: those that bear crops above ground should be planted during the light of the moon, while those that bear crops below ground should be planted during the dark of the moon, after the full moon has passed. The moon-phase information harkens back to the eighteenth and nineteenth centuries, when it was commonly believed that the moon and the planets influenced plant growth.

Then come the famous weather forecasts: daily weather predictions for sixteen regions of the United States, all composed unflinchingly a full eighteen months in advance. Although some may snort at the sight of the yellow cover, the almanac's forecasts do represent a serious attempt at long-range forecasting, the only such forecast available. From 1970 to 2002, the almanac employed a solar physicist and former NASA scientist by the name of Richard Head who reportedly "enhanced" Thomas's secret formula by factoring in solar activity, in particular sunspot cycles, and other variables such as the El Niño currents. Since 1995, meteorologist Michael Steinberg has been making the weather predictions, using climatology and meteorology in addition to solar science. Traditionally the editors of the almanac have insisted that the forecasts are 80 percent accurate (some weather researchers would beg to differ).

Say what you will, there have been some uncanny predictions. The almanac predicted the devastating freak tornado that hit Worcester, Massachusetts, in 1953 with the warning, "Heavy squall and that's not all." Hurricane Andrew of 1992 was forecast almost perfectly. Perhaps the most famous prediction ever offered by the almanac was for the summer of 1816. Legend has it that while the founder Robert B. Thomas was down with the flu, the printer inserted a prediction for snow in July and August. When Thomas discovered the forecast he was not amused and set about correcting as many copies as he could lay his hands on. When it arrived the summer of 1816 brought winter conditions and heavy snowstorms to New England, most likely due to a volcanic eruption in the Dutch East Indies. Needless to say, the coincidence did not hurt the almanac's reputation. Nor did an incident that occurred during World War II in which the FBI captured a German spy who had been landed by a U-boat on the shore of Long Island. In the spy's coat pocket was *The Old Farmer's Almanac.*

Today the *Almanac* still heeds the words of its founder: "Our main endeavour is to be useful, but with a pleasant degree of humour." And Thomas's secret formula to this day rests safety in a black tin box in Dublin, New Hampshire.

The Old Farmer's Almanac is released every year on the second Tuesday of September. To learn more about *The Old Farmer's Almanac* or its family of publications, including the *Gardener's Companion* magazine and an array of calendars and journals on weather, gardening, herbs, and more, visit the website www.almanac.com. *The Old Farmer's Almanac* and Old Farmer's Almanac products are published by Yankee Publishing Inc. of Dublin, New Hampshire 03444.

GARDENING MOVEMENTS

Theophrastus's Roses

The greatest plant classifier was Carl Linnaeus (1707–78), but the initial steps in this field were taken some two thousand years earlier by an often overlooked Greek philosopher. What was so revolutionary about Theophrastus's approach was that it was systematic—no one before had ever thought to classify a rose by counting the number of its petals.

Theophrastus was born the son of a cloth worker at Eresus, in Lesbos, in about 371 B.C. He went to the Athens Academy to study under Plato, from whom he learned the importance of systematic thought and attention to detail. Most significantly he learned that in order to study things it was necessary to classify them. It was presumably at this time that he became acquainted with Aristotle. When Plato died, Aristotle founded the famous Lyceum, known as the "Peripatetic School" because the teachers used to lecture while strolling in its groves and gardens. As an educational center, it was equivalent to a university. Appointed as Aristotle's deputy, Theophrastus was given responsibility for the gardens, and in that capacity was allowed to travel widely abroad searching for plants. When Aristotle was forced to leave Athens in about 322 B.C. Theophrastus was appointed his successor. After Aristotle's death he inherited the old philosopher's library and unpublished writings, which he continued to work on. He also became head of the school, which he enlarged, continuing as its leader for thirty-five years.

Theophrastus was ideally qualified to become a gardening guru. The philosopher, writer, and naturalist was well able to apply his multi-talented skills to his knowledge of plants, largely derived from his earlier travels. In addition to his own wide knowledge, he was well supplied with horticultural data by farmers, woodsmen, and charcoal-burners, as well as by the students and staff at the Lyceum. Alexander the Great, while engaged on military expeditions in the Levant, also sent him many plants. Arguably Theophrastus's large and systematically arranged garden at the Lyceum was the world's first botanic garden. Attention to detail and a propensity toward empiricism, allied to contemplation, were the hallmarks of the school which Theophrastus promoted. He realized that his disciplined principles, the work of the garden, and his own interest in plants mattered at a time when there were plenty of irresponsible plant writers in circulation. The latter were known as rhizomatists—quack doctors who insisted that almost any plant had curative powers.

Thus there was every reason for Theophrastus's book *An Enquiry into Plants* to break new ground. And it did so. It contained a systematic attempt to classify all known plants, the first in western literature. Noting that wild plants were largely unknown and untamed he concentrated most of his efforts on plants that were cultivated. Even then he found himself so constrained by imprecise terminology that he needed to introduce several new technical terms. There, and in the subsequent book *The Causes of Plants*, Theophrastus wrote about the first roses known to the Greeks. He revealed that there were several kinds, differing in the number of their petals—ranging from five to a hundred. They differed, too, in their color and sweetness of scent, and not all were fragrant. Another interesting fact was that red roses were unknown to Theophrastus—only white and pink varieties were grown in Greece.

Theophrastus died in 288 B.C., leaving his priceless garden to the school. He was largely overshadowed by Plato and Aristotle until 1916, when Sir Arthur Holt, a distinguished Greek scholar and also a learned amateur gardener, wrote a readable translation of *An Enquiry into Plants*, and the significance of his achievements came to be properly understood. Theophrastus was probably the greatest botanical writer before the Renaissance.

 Useful website: www. gardendigest.com.

Alexandre Dumas and the Camellia Craze

This story is a unique example of how a dramatic novel—by a nongardener—fueled a sensational mass enthusiasm for a flower. Such was the influence of Alexandre Dumas's *La Dame aux Camélias*.

A relative of the tea plant, the camellia is a large evergreen shrub that can grow to a height of 30 feet. It has thick, dark, glossy foliage and bowl-shaped flowers with overlapping petals ranging from white to pink and red. Linnaeus the great botanist had named it after George Joseph Camellus (or Kamel), a Moravian Jesuit who traveled in Asia and wrote an account of these plants which he discovered growing on the Philippine island of Luzon. The leaves of the plant were made into decorations and worn by Japanese women in their hair. Many of the numerous cultivated species originated from India, China, and Japan and were introduced into Europe by Lord Petre in 1739. By the mid-nineteenth century it was gaining in popularity in Europe, but Dumas was to change all that.

Alexandre Dumas ("Dumas *fils*") was born in Paris on July 27, 1824, the natural son of the playwright Alexandre Dumas. Initially not acknowledged by his father, the child lived for his first five years in poverty with his mother, then was sent at his father's expense to various schools in Paris. At the age of seventeen he finished his formal education and he was let loose on the Parisian world without having to choose a profession or do anything but enjoy the facile pleasures which Paris afforded.

One of his numerous affairs was with the young courtesan Marie du Plessis. Alexandre was not his father's son for nothing. Although he sowed, sometimes in his father's company, a plentiful crop of wild oats, the knowledge he gained of the seamy side of life was not wasted. Then suddenly the lad who had been taught to regard his prodigal father's resources as inexhaustible was rudely awakened. The coffers were empty and he had accumulated debts to the amount of two thousand pounds—a vast sum. Desperate for money, he pulled himself together. To a son of Dumas the use of the pen came naturally and he embarked on his writing career.

The outcome, written at terrific speed, was the novel *La Dame aux Camélias* (1848), which he based on his affair with

Marie du Plessis. The fictional heroine, Marguerite "Camille" Gauhier, was a beautiful high class courtesan who became part of the fashionable world of Paris. Scorning the wealthy Count de Varville who offered to relieve her debts should she once more become his mistress, she escaped to the country with her penniless lover Armand Duval. There Camille made a great sacrifice. Giving Armand, whom she truly loved, the impression that she has tired of their life together, at the request of his family she returned to Paris and her life of frivolity. The tale concluded with the tragic reunion of Armand and the dying Camille.

Significantly, Marguerite indicated her "availability" by wearing white camellia blossoms for twenty-five days of the month, and red for the other five. The novel caused a tremendous scandal, and a clash with the censors delayed its conversion into a popular play. Nevertheless the public was hooked. The play (1852) was a phenomenal success, becoming the most viewed production in the Second Empire. The effect it had on the sales of camellias across Europe, and also the United States, was quite amazing. Originally the plants had arrived in the United States in the late eighteenth century and soon became popular among gardeners. Now new plants were snatched almost from under the noses of nurserymen. A hungry public cried out for more and more varieties. The demand was further intensified when *La Traviatta*, Verdi's opera based on the novel, was premiered in Venice in 1853.

One factor fueling the craze was Dumas's careful choice of the camellia. Emphysema, an incurable and fatal coughing disease, was rife and killed hundreds of thousands of people in Europe and America each year. Marguerite, with her live-fast-die-young attitude, was recognized as suffering from this affliction, and the reason why she carried only unscented camellias was because all other blooms made her cough.

The camellia phenomenon established the younger Alexandre Dumas's reputation as a major author and playwright, and he went on to accumulate a fortune before his death in 1895. Such has been the enduring appeal of his story that well-wishers still visit both his grave and Marie's. Due to the immense popularity of the novel, the plants themselves became so commonplace that they ceased to be favored by the aristocracy. The withdrawal of upper class patronage meant that by about 1870 the flower had lost its popular fashionable flavor and although the story's appeal endured, the camellia craze came to an end.

 For more details on camellias, contact the International Camellia Society, www.med-iz.iuni-sb.de.

🌺 Titus Salt's Garden Plots

Readily identifiable by their higgledy-piggledy rows of vegetables, runner beans, and ramshackle potting sheds, garden plots made practical gardening accessible to millions of working class urban Britons, particularly in wartime. The influential pioneer who did most to facilitate small plot–gardening as a recreational activity was a philanthropic Victorian mill-owner uniquely blessed with daring entrepreneurialism and a compassionate perception of human nature.

Sir Titus Salt created his fortune by discovering a method of processing alpaca, the lustrous dark fleece of South American camels. Born in 1803, the son of a Bradford wool-stapler, by the age of thirty-three Titus had successfully devised a means of untangling coarse wool to manufacture worsted and was running four mills in Bradford. In 1836, while inspecting wool in a Liverpool warehouse, he happened to come across the first imported cargo of alpaca hair, widely ignored as it was thought to be too long to be workable. Buying the entire consignment he successfully adapted his worsted machinery to turn alpaca into a quality textile, and from there on cornered the market. To boost production of his alpaca textiles he began building the largest mill complex in Europe by the River Aire near Bradford in the early 1850s. In the monumentally extravagant construction of Saltaire no expense was spared, and he provided state-of-the-art facilities for all the workers. There was an infirmary, a park, almshouses, and eight hundred model dwelling houses— every one of which was entitled to a garden plot.

Each cultivable plot measured some 30 by 89 feet: "an area not exceeding 15 poles in extent." For eager Saltaire

The philanthropist Sir Titus Salt treated his mill-workers unusually well. (Bradford Libraries)

plot-holders the nature of small garden cultivation made it essential to adopt some horticultural principles: row cultivation was almost universal, with the rows neatly laid out and meticulously tended, with due attention paid to a three-year rotation of crops. Typically there might be winter and summer crops of potato, lettuce, carrot, cabbage, beet, dwarf peas and beans, as well as tomato, squash, radish, parsley, rhubarb, leek, onion, shallot, parsnip, broccoli, kale, and sprouts. In a corner there might be a seed-bed guarded by a scarecrow. For housing tools and storing crops on each patch there was usually a shed commanding pride of place as an expression of individuality. Most were rickety, perhaps made from old window frames or fence panels, with a roof of tarpaulin (later corrugated iron).

Salt believed small gardens to be morally enhancing in that plot-holders derived therapeutic delight from growing their own crops, while at a practical level workers who slaved in

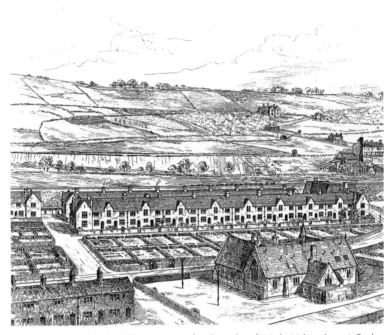

The Saltaire plots influenced the provision of such gardens for industrial workers at Copley village. (*The Homes of the Working Classes*, 1866)

the satanic mills all day could relax in the fresh air. If his employees spent their spare time in useful toil on the gardens, he reckoned, it would also keep them away from the temptations of the pawnshop and pub. A devout Congregationalist, he encouraged plot-holders to attend church and not work the land on Sundays. He even strategically sited the Saltaire plots opposite the mill, between the park and the church (in 1966 the whole complex was listed as a Conservation Area).

The notion of escape into another world was a huge attraction for all plot-holders, the garden often representing a haven where solace could be found. On his plot the holder was in control. His life became centered around how he treated the ground and the crops he grew. For many poor families they were a *rus in urbe*, ideal for picnicking and gossip. Cultivation became the basis of a social culture enjoyed by most plot-holders, often in the form of healthy competition among them as to who should have the earliest vegetables of each kind. There was also cooperation; this began with the sharing of compost and fertilizers, and grew into the formation of garden associations. Eventually in 1930 the establishment of a National Allotments Society enabled plot-holders to bulk-buy seeds and also tools and insurance at preferential rates.

Ambassadors, prime ministers, and even royalty came to view Saltaire as its renown spread. It was described as the world's healthiest manufacturing town. Salt was rewarded for his enlightened philanthropy with a knighthood, and a statue of him was erected in Saltaire Park. Inspired by his grand plots, eminent mining, railway, gas, and canal magnates gradually began to provide their employees with urban plots.

Yet in rural areas, of the 604,250 acres of common land fenced off by thousands of separate Enclosure Acts between 1845 and 1869 only 2,175 acres were assigned to the distressed. Cumulatively the Enclosures had the effect of depriving a great mass of landless poor of the right to common grazing that they had enjoyed for centuries. Insofar as there had been any provision of plots in rural areas it had taken the form of occasional scraps of land set aside by a few aristocrats alarmed by the decline of the agricultural workforce. Rural parish councils were not required by central government to set aside land for garden plots, and they invariably prevaricated as they regarded such plots as shanty towns. Yet they were usually happy to rent out to the middle classes plot-sized leisure "guinea" gardens.

Salt was elected to Westminster in 1859 at a time when associations were busily banding together to form the Allotment Movement—a nationwide working-class group agitating for council garden plots. Eventually an Allotments Party was formed. This proved immensely popular, and when it won control over Lincolnshire County Council Parliament was jolted into action. In 1887 the Allotments Act was approved, requiring (instead of only encouraging) councils to provide land for such plots. With the political battle fought and won, enthusiasm for small plots strengthened. Between 1873 and 1914 the number of plots in Britain more than doubled from 245,000 to 600,000.

During World War I the main purpose of these small gardens swung from recreation to functionality. Special powers were granted to requisition land, and the number of plots increased to 1.5 million. The social significance of the creation of this vast army of shirt-sleeved volunteers to cultivate those plots was that the war spread gardening far beyond existing limits of class or occupation, familiarizing the wider population with the habits and techniques of vegetable-growing.

During World War II the "Every-Man-a-Gardener" campaign was revived as "Digging for Victory." With the country facing the real danger of semi-starvation, plot-holders achieved absolute miracles of production, supplying an estimated 10 percent of all food available in Britain. Necessity meant that every available urban green space was dug up and sown for vegetables. Even Saltaire Park, around the statue of Salt—who had died in 1876—was transformed into makeshift plots. Titus Salt would doubtless have been pleased.

The American Community Gardening Association can be contacted at www.communitygarden.org. Other useful websites are www.gardenclub.org and www.cityfarmer.org.

Mary Montagu's Language of Flowers

This is a bizarre story of how the respectable name of an important medical pioneer came to be trivialized by being unfairly and cynically associated with a plant movement. Since ancient times symbolic meanings had often been attached to flowers, especially in religion, heraldry, painting, and literature. There were also plant-related superstitions, sacred tree mythologies, and other beliefs, but these were mere folklore and generally lacked intellectual respectability. Then in 1868 John Ingrams wrote and published *Flora Symbolica*, in which he claimed all that had changed. He was a respected writer, who had produced a biography of Edgar Allen Poe. Certainly he was right in some respects, as in the 1860s there had been a phenomenal growth of

Lady Mary Wortley Montagu was erroneously made a figurehead of the "language of flowers" movement. (Gibbs, *The Admirable Lady Mary*, 1949)

interest in what he termed the "language of flowers," which had all of a sudden become minutely codified. Numerous books appeared, most written by women, containing lists of hundreds of plants and flowers and giving their meanings. At the huge and ordinarily conservatively stocked library at the Royal Botanic Gardens, Kew, there are literally dozens of such books from that era covering several shelves. Controversially, Ingrams claimed in his book that the public figure most responsible for this phenomenon was Lady Mary Wortley Montagu.

Born in 1689, the eldest daughter of the Earl of Kingston, Mary showed early literary abilities in learning Greek and Latin, although she also enjoyed the privileges of an aristocratic upbringing. In 1712 she married Edward Wortley Montagu MP, and she accompanied him to Constantinople when he was appointed British Ambassador there in 1716. Tragedy soon struck. Mary, a great society beauty, caught smallpox, which devastated her good looks. Fearing for the health of her son, she heard of a local custom of inoculating against smallpox and determined to try it. Finding that it worked, she used her influence to introduce smallpox inoculations into England. Renowned for her verbal wit and written poetry during her lifetime, she became immortalized a year after her death in 1762 when her correspondence was published as *Letters from Turkey*.

Ingrams, and later other writers, claimed that this book had, many years later, ignited the language of flowers boom. In fact close inspection of the text shows very few mentions of plants or flowers, although Lady Mary did refer to the custom of courtesans in Turkish harems using flowers and other artifacts to communicate with their lovers outside the walls. Essentially the unscrupulous Ingrams had made use of

L E T. XL.

TO THE LADY——

Pera, March 16. O. S.

I AM extremely pleased, my dear lady, that you have, at length, found a commission for me, that I can answer, without disappointing your expectations; though I must tell you, that it is not so easy as perhaps you think it; and that if my curiosity had not been more diligent than any other stranger's has ever yet been, I must have answered you with an excuse, as, I was forced to do, when you desired me to buy you a Greek slave. I have got for you, as you desire, a Turkish love-letter, which I have put into a little box, and ordered the captain of the Smyrniote to deliver it to you with this letter. The translation of it is literally as follows: The first piece you should pull out of the purse, is a little pearl, which is in Turkish called *Ingi*, and must be understood in this manner:

Ingi, *Pearl,*	Senfin Uzellerin gingi *Fairest of the young.*

Caremfil, *Clove,*	Caremfilfen eararen yok Conge gulfum timarin yok Benfeny chok than feverim Senin benden, haberin yok.

> *You are as slender as the clove!*
> *You are an unblown rose!*
> *I have long loved you, and you have not known it!*

Ful, *Jonquil,*	Derdime derman bul *Have pity on my passion!*

Kihat, *Paper,*	Birlerum fahat fahat *I faint every hour!*

Ermus,

A letter sent by Lady Mary from Turkey was claimed to have sparked off the movement. (*Letters of Lady Mary Montagu*)

the good name of the ambassador's wife to provide the absurd language of flowers movement with an element of respectability. Lady Mary's dying words summarized it well: "It has all been most interesting."

 Visit the ancestral home of the Montagus, Boughton House, near Kettering, Northamptonshire; see also www.newcriterion.com.

Eve Balfour, the Soil Association Guru

The individual who did most to pioneer the modern organic gardening movement was Lady Eve Balfour. Her influential Haughley Farm, and her book *The Living Soil*, showed how care for the soil could affect human health.

Headstrong and altruistic, from an early age Lady Eve knew she wanted to be a farmer, an ambition she was likely to attain since she came from a powerful family of high achievers. Her father Gerald Balfour was the Chief Secretary for Ireland, her maternal grandfather was the Viceroy of India, and her uncle A. J. Balfour was prime minister from 1902 to 1906. Her passion for the great outdoors was perhaps inherited from Arthur Balfour's younger brother Francis, an eminent biologist and alpine plant enthusiast, who had been tragically killed during an ascent of Mont Blanc's unconquered Aiguille Blanche.

In 1915, at the age of seventeen, Eve became one of the youngest women to study agriculture at the University of Reading. She soon gained practical experience, organizing Land Army girls on a farm in World War I. Then in 1919, when aged only twenty-one, she bought her first farm in Stowmarket, Suffolk. She weathered the arduous working life with Bohemian panache. In her spare time she successfully co-authored three detective novels. Already a talented flute player, she took up the saxophone and played in a jazz band with her sister and friends who lived at the farm. Their gigs included Saturday night dances at a hotel in nearby Ipswich.

She cut her campaigning teeth by successfully fighting against the unpopular tithe tax in the 1930s, but it was only in 1938, when she was forty, that Eve Balfour read the book that was to profoundly alter the direction of her life. Viscount Lymington's *Famine in England* not only threw into doubt assumptions about the sustainability of orthodox agriculture, it also indicated there could be alternative strategies. Eager to know more, Eve began devouring every publication she could get her hands on concerning that and related subjects. In the process she encountered the work of Sir Robert McCarrison, a medical doctor who was director of a nutritional research survey in India. In a study of the remarkably long-living Hunza tribesmen on India's northwest frontier, McCarrison found an apparent correlation between their agricultural practices and their exceptionally robust health. Their diet was entirely wholefood, grown on humus-rich soil fertilized by composting. Eve also learned that Sir Albert Howard, a distinguished economic botanist also working in India, had separately observed that traditional eastern methods of composting yielded fodder which better enabled animals to resist disease. Howard had come to realize that organic agriculture was the best form of preventive medicine.

Balfour allegedly switched her own diet to one based totally on whole and unprocessed foods. Aware of the inadequacies of western knowledge about organic husbandry, she decided to embark on a closely monitored study of contrasting land use. By 1939 she had developed her own agricultural premises at Haughley in Suffolk, and she now decided to turn the farm there into an experimental center. Renamed the Haughley Research Trust, the farm was divided into comparable units, the most significant of which were the mixed section, on which some chemical fertilizers were used, as well as herbicides, insecticides, and fungicides when thought necessary, and the organic section, where no chemicals were used. During the war years Eve ran the farm short-staffed, and gathered together an eclectic assortment of evidence to support her theories from subject areas as diverse as medicine, nutrition, agriculture, and environment, not just from Britain but from elsewhere in the world.

The result was the publication in 1943 of *The Living Soil*— effectively the first book on organic farming. In the introduction she declared: "My subject is food, which concerns everyone; it is health, which concerns everyone; it is soil, which concerns everyone though they may not realize it—and it is the history of certain researches linking these three vital subjects." An outstanding intellectual achievement for someone who was not a professional scientist, the work was an enlightened synthesis of the scarcely comprehensible works of medical, agricultural, and environmental scientists. The timing of her multi-disciplinary, holistic work was perfect: the population's general physical condition in the 1930s had been poor; there was a case for a health policy founded on preventive medicine; and a National Health Service was being disputed. Anxiously, however, the government was encouraging an "industrial" approach to farming and was even subsidizing the use of agricultural chemicals.

The Living Soil had never been expected to attract a mainstream readership, but it was a bestseller and within five years it was into its eighth edition. In response to the enormous interest in the book, a meeting of 160 far-sighted sympathizers was convened in London in June 1945. This resulted in the formation of the Soil Association the following year, and this was by no means the haphazard grouping it has commonly been depicted as. Using her political skills, Lady Eve hand-picked the Association's founder-members, choosing individuals she reckoned might be influential in promoting its aims of bringing together all those working for an understanding of the vital relationship between soils, plants, animals, and humanity; assisting research in that field; and distributing the knowledge gained in order to create a body of informed opinion.

Standing at the forefront of opposition to prevalent western agricultural practices, Balfour and her followers attracted much criticism, especially from farmers. The organic advocates were often dismissed as nostalgic "hobby farmers," elitists with precariously uneconomic ideals. Some detractors tried to link them with the teachings of Rudolph Steiner, an Austrian philosopher who in 1929 had established the Bio-Dynamic Agricultural Association, which was a rather more mystical organization than the Soil Association insofar as it advocated belief in the power of lunar cycles and cosmic forces. More difficult to refute were allegations that the trial results at Haughley, now under Soil Association control, were less clear than had been hoped. Most damningly Albert Howard never joined the Association because he was dubious about the scientific validity of the Haughley trials. Regardless of the detractors, in time the organic movement grew. Balfour's influence inspired the publication in America in 1963 of Rachel Carson's *Silent Spring*, which greatly increased public awareness of the natural environment and warned against the dangers of pollution.

Gardening had been predominantly organic until the 1920s and it started to be so again from the late 1950s onward, largely through the efforts of the gardening journalist Lawrence Hills. A founding member of the Soil Association, in 1958 he established the Henry Doubleday Research Association (HDRA), which he named after an innovative nineteenth-century Quaker smallholder. Hills inspired many people to take up organic gardening, and for years acknowledged the influence of Eve Balfour by having HDRA adopt Soil Association guidelines of what was, and was not, organic. Her groundbreaking work provided a framework whereby amateur horticulturists could apply their concerns about healthy eating and the overuse of chemicals to their own gardens.

Beyond the world of agriculture Lady Eve had other interests in addition to her jazz band. She was a qualified pilot and an experienced yachtswoman who regularly crewed on her brother's yearly voyages to Scandinavia. Though dedicated to the organic cause she was no zealot. Before she died in 1990 someone asked her about her own diet. "Oh, don't model yourself on me," she replied. "I drink gin and tonic and smoke cigarettes. So long as you're good 75 percent of the time, the remaining 25 percent will look after itself!"

For more on organic gardening, see www.organicgardening.com, www.hdra.org.uk, or www.soilandhealth.org.

Nurseries

John Gerard the Herbalist

A barber-surgeon who amused readers of his influential book *Herball* with weird notions such as a Barnacle Tree which produced live geese, John Gerard was nevertheless the best known of all English herbalists. He was born in 1545 at Nantwich, Cheshire, educated at a grammar school in Willaston, and in 1562 was apprenticed to an eminent barber-surgeon in London. On completion of his seven-year apprenticeship, he became ship's surgeon on a merchant vessel trading in the Baltic. He visited Russia, Sweden, Poland, and Denmark, and may even have sailed to the Mediterranean. Settling in London, in 1569 he was admitted to the freedom of the Barber-Surgeons Company, and established his own medical practice in Holborn. He is believed to have concentrated on the medical, rather than the haircutting, elements of his profession.

While serving his apprenticeship, Gerard's interest in plants developed into a passion. He started a nursery garden near his cottage on the corner of Fetter Lane in Holborn, which was then a fashionable part of London. In his leisure time he

Portrait of John Gerard, from his *Herball*. (By courtesy of the National Portrait Gallery)

worked hard there, reportedly "labouring with the soil to make it fit for plants, and with the plants, that they might delight in the soil." An enthusiastic botanist, he searched for native plants near the capital and further afield in Margate, Rye, Harwich, and Cambridge. Gaining some renown as a "gardening doctor," he was often presented with rare plants and seeds from different parts of the world and he also received a number of job offers to supervise the gardens of noblemen. He was made superintendent of the gardens of William Cecil, Lord Burghley, both in the Strand and at Theobalds in Hertfordshire, a post which he held for twenty-one years.

In 1596 he compiled a list of the plants cultivated in his garden. The *Catalogus* was unique insofar as it was the first fully comprehensive catalog ever published of any one garden. It listed over a thousand species and indicated that many of the plants on his patch were new introductions to England. George Baker, Queen Elizabeth's surgeon, observed that the garden had "all manner of strange trees, herbs, roots, plants, flowers and other such rare things." A second and expanded edition, dedicated to Sir Walter Raleigh, followed three years later.

Gerard's reputation principally rested on the *Herball, or Generall Historie of Plantes* (1597), which made his name a household word. The book consisted of more than 1,800 woodcuts, including the first printed illustration of the potato. Entertainingly there were many of Gerard's own remarks, such as directions for finding scarce plants in various parts of England, and also allusions to people and places of high antiquarian interest. Yet although Gerard took complete credit for the work, the *Herball* was not quite as original as he made out. In fact, it was almost entirely based

Part of the frontispiece of the *Herball* showing herb beds.

on a translation of *Stirpium Historiae Pemptadades Sex* (1583) by the Flemish botanist Rembertus Dodoens. Only sixteen of the woodcuts were done by Gerard, the rest came from the blocks used by Jacob Theodorus Tabernaemontanus's *Eicones Plantarum seu Stirpium* (1590). Essentially Gerard had taken over Robert Priest's unfinished translation of Dodoens's last book and enhanced it with his own plant searches and garden experiences.

Nevertheless, the first edition of the *Herball* was immensely popular and it led the field unchallenged for more than a generation. It provided a large amount of folklore, such as the imaginary Barnacle Tree—Gerard claimed to have observed geese hatching from the tree's fruit! Yet with it was valuable information on habitats, time of flowering, and uses of plants, all of which contributed to the book's success. It also influenced the thinking of later botanists and poets, including Milton. A timely book, the *Herball* was itself effectively a watershed in the transformation of public perceptions of the purpose of gardens, from medicinal storehouses to horticultural pleasure grounds.

In addition to running his own nursery Gerard became the curator of the garden established by the College of Physicians in the City of London. This did not evolve quite as he had hoped. His ambition was to develop it to such an extent that barber-surgeons could grow medicinal plants there and hence become thoroughly acquainted with them. Had his plan been put into practice he would arguably have achieved the great honor of having created the first botanical garden in Britain. However, he was named Master Barber-Surgeon, and thereby the chief examiner of surgeons practicing in London. In 1612 he died and was buried in St. Andrew's Church, Holborn; despite his accomplishments, there was no monument to mark the spot.

 For further information on herbal medicines, contact the Society of Apothecaries of London, Blackfriars Lane, London EC4V 6EJ; www.apothecaries.org.

 John Tradescant's "Ark"

Spectacularly furnished with many of the plants gathered by John Tradescant the elder when he traveled abroad as an adventurous envoy, the "Ark" was a notably famous early commercial nursery.

John Tradescant was born at Corton on the Suffolk coast in 1570, but little is known of his early life. In 1607 he married

Tradescants senior and junior. (By courtesy of the National Portrait Gallery)

a Kentish girl at Meopham, producing a son whom they named John. The couple had met while Tradescant was working as the supervisory gardener at Cobham Hall, home of the statesman Robert Cecil, Earl of Salisbury, who was busy at the time constructing a grander home at Hatfield House. Cecil's regard for Tradescant's intelligence, shrewdness of negotiation, and integrity was such that he was dispatched by Salisbury on a shopping expedition to Holland, Flanders, and France to purchase the very best fruit trees, shrubs, and plants for the new garden.

After Cecil's death Tradescant became gardener to Lord Wotton at Canterbury, and it was probably through Wotton that he chanced to acquire shares in an expedition to develop the new colony of Virginia, in the hope that members of the expedition would bring him new plants they found there. In 1618 Tradescant succeeded in embarking on a trade mission to Russia, traveling as the mission's naturalist. At Archangel he astonished the Muscovites by his enthusiasm for plants that they considered to be weeds. He returned home with a priceless cargo of plants and specimens, including angelica and the Siberian larch. His notebook, brimming with his descriptions of flora, was the first known register of Russian plants. Next, this time as a "gentleman volunteer," he joined an expedition against Algerian corsairs in the Mediterranean. At considerable risk to himself he made several trips ashore. He returned with the Algiers apricot, which he propagated; eventually one was to be found in every nobleman's garden.

A particular characteristic of Tradescant's was that—unusually for a gardener—he had no sense of smell. This may have been a factor in his growing urge to seek artifacts that were strange and rare. In Russia he discovered the abacus, which was then the chief calculating machine, and brought one home: It was the first ever seen in England. Tradescant then moved on to work for the Duke of Buckingham, himself an obsessive collector and the most powerful figure at court. For his new employer he was not just a gardening adviser but also a special envoy, and he was assigned to escort Henrietta Maria, the bride of Charles I, from Paris to England. Significantly Tradescant was made principal accumulator of a scheme to create a highly prestigious collection of "rarities." He was now being highly paid to indulge his eccentricity. Subsequently, in the course of his foreign travels, he gathered an extensive range of unusual objects such as elephants' heads, as well as seeds, plants, trees, and shrubs—and the systematic Tradescant was ever careful to collect duplicates.

When Buckingham was murdered, Tradescant became the royal gardener. Now financially independent, he began to establish his own botanical garden. In order to be near likely sources of patronage and fellow gardening professionals he settled in a graceful-looking house with a few acres of disused farmland by the Thames at Lambeth. In addition to his own trees and shrubs there were some seven hundred varieties donated to him from Buckingham's various gardens, while previously unseen species also arrived from Virginia, sent by John the younger, Tradescant's son. His introductions from the New World included the Virginia creeper. But the Tradescants were not solely concerned with the exotic and colorful, and the pioneering nursery included many less glamorous flowering plants. They printed a comprehensive catalog of the plants for sale in 1630 and thereby provided ordinary working gardeners, as well as society figures, with a greater opportunity to view and acquire these rare new finds.

An astute operator, Tradescant the elder knew that nurseries were largely dependent for their trade on novelty. He was fortunate to have associations with royalty. But what really pulled in the crowds, and gave the premises their nickname the "Ark," was the growing collection of duplicate rarities on display in the house. Among the more eccentric exhibits alleged to be there were a dodo and a dragon's egg. Effectively the Ark became England's first museum open to the public. It must have been the most exciting house in the world to live in, as it welcomed a steady stream of seafarers, naval commanders, and merchants, just returned from voyages to India, China, or the Americas and bearing strange new plants and fruits.

Oliver Cromwell never took umbrage against the remarkable premises at Lambeth, and so the Ark was able to remain in business uninterruptedly during the Commonwealth. Thus the museum's curios, which were deemed to be of educational value, continued to attract exclamation and congratulations. The utilitarian nature of the exotic fruits and vegetables far outweighed other considerations. In those troubled years the nursery became the Tradescants' main source of income.

The prospect that the Tradescants might become a dynasty of nurserymen came to an abrupt end when Tradescant the elder's grandson died before his son. A few years after Tradescant the elder's death in 1638 the Ark's artifacts were sent to an antiquarian named Elias Ashmole, who in turn transferred them to Oxford—where they became the founding collection of the Ashmolean Museum. Many of the nursery plants went to the Oxford Physic Garden. The elder and younger Tradescants were both buried at Lambeth. Poignantly their tomb lies in the church of St. Mary-at-Lambeth which has since become the famous Museum of Garden History.

 The Tradescant Gardeners of America can be contacted via The Museum of Garden History, Lambeth Palace Road, London SE1 7LB; www.museumgardenhistory.org. For information on Oxford's botanic gardens, see www.parks.ox.ac.uk.

Thomas Fairchild's "City Gardens"

One of the most influential nurserymen in the history of gardening, and perhaps the greatest florist of all time, was the botanist and entrepreneur Thomas Fairchild. In 1717, while doing experiments to investigate the relevance of sex in plant reproduction, he successfully fertilized the style of a carnation pink with the pollen of a sweet william. He thereby became the first person to raise a hybrid scientifically. Whereas grafting had been done for centuries, this hybridization, *Dianthus caryophyllus barbatus*, was revolutionary. When he unveiled it to the Royal Society it caused an uproar and was dubbed "Fairchild's Mule." A man of considerable religious faith, Fairchild had serious moral doubts about the "Mule." Unlike many later scientists, he wondered whether his production of new forms, albeit by the artificial mixing of those already in existence, might somehow be an unscrupulous manipulation of nature.

So highly regarded was Fairchild's scientific knowledge that he was invited to present papers to the Royal Society on other botanical subjects. Remarkably, however, there seems to be no evidence that he ever received any formal education in science. He was born in 1667 in Cripplegate, London, but virtually nothing is known of his early life; later he was apprenticed first to a clothworker before switching his apprenticeship to gardening. By 1690, at the age of twenty-three, he was already the proprietor of his own nursery in Hoxton, near the City of London's village suburb of Shoreditch—clearly visible from the dome of Wren's new St. Paul's Cathedral.

Called the "City Gardens," the nursery covered just half an acre of ground, and it was crammed with hot-beds and greenhouses where Fairchild conducted his botanical experiments. He also cultivated one of England's few vineyards, from which he sold fifty varieties of grape. The nursery's success was undoubtedly partly due to Fairchild's immense energy and business acumen. Moreover, his innate technical know-how was combined with a quite uncanny natural ability to grow almost any plant. As the principal cultivator of exotics at the City Gardens, he introduced his customers to a wide range of fine North American plants, many of which were supplied to him in seed form by the plant finder Mark Catesby. Fairchild's famous collection of aloes and succulent plants was unsurpassed in the country, as was his fruit garden, which was described by the writer Richard Bradley in 1721 as "the greatest collection of fruits I have ever seen, and so regularly disposed that I do not know any person to excel him in that particular."

The increasing ownership and use of small private gardens in the early eighteenth century was poised to make gardening a popular recreational amenity for London's urban middle classes. Yet more often than not plants purchased from nurseries perished because they were wrongly planted, or just inappropriate for their new conditions. The heart of London was still a dirty, unpleasant place in which to live, and it was so polluted it was amazing that anything grew there at all. Vast quantities of coal were burned domestically and industrial premises produced great clouds of smoke that enveloped the city like a mantle. The black pall of smog that hung everywhere was one of the characteristics of London, and in 1661 the celebrated Deptford diarist John Evelyn had become so enraged by the unhealthy environment that he produced a book on the subject, *Fumifugium: or the Inconvenience of the Air and Smoke of London*. He grumbled: "It is the horrid smoke which kills our bees and flowers, suffering nothing in our gardens to bud, display themselves, or ripen."

Having been persuaded by Richard Bradley to write a book on the subject of gardening in adverse conditions, Fairchild put his ideas and experiences down on paper. The result was *The City Gardener*, a delightfully unpretentious 70-page gem crammed with information of a very varied character. Published in 1722, it was the first book ever to tackle the specific problems of town gardens. In a lively style, Fairchild insisted that much could be done with existing urban plants, and he advocated the planting of trees and shrubs such as the box privet, holly, ivy, lilac, laburnum, and passion flower. He showed how fashionable squares could be improved with an appropriate choice of trees, and reassuringly he cited many obscure nooks and crannies in the city where he had seen such examples flourishing. Furthermore, he insisted, courtyards and narrow streets with restricted light could easily sustain exotic architectural plants.

Fairchild took a dim view of flower markets, ruefully noting that he had seen in them plants "which are uncertain of growth as a piece of Noah's Ark," though he conceded: "I suppose this chiefly happens through the ignorance of the sellers of plants." As for those Londoners who wanted to decorate their balconies, he recommended the growing of small fruit trees in tubs. Despite his apparent altruism he was ever the entrepreneur and somehow could not resist noting that all the items mentioned were available at his Hoxton nursery!

So many people were acquainted with London's lifestyles, markets, and trends that the capital exercised a considerable influence over national taste and consumer behavior. Thus Fairchild's pioneering book, which established town gardening as a viable leisure activity by means of showing the ideal plants for urban survival, also influenced other nurseries. It inspired seedsmen, nurserymen, and horticulturists to cater to the specialized requirements of the urban bourgeoisie. When the book was reprinted, anonymously, in 1737 under the title *The London Gardener*, Fairchild's influence on urban gardening had spread far and wide to other Georgian cities, notably Bath, Dublin, and Edinburgh. The latter's smoky atmosphere had won it the nickname "Auld Reekie."

The gregarious Fairchild was a founder member of the Society of Gardeners—a leading group of commercial horticulturists

The frontispiece to the Society of Gardeners' *Catalogus Plantarum*.

who met every month in a Chelsea coffee house to discuss the latest plants which members brought along from their own nurseries. Considered by his contemporaries to be the pre-eminent—and certainly most famous—nurseryman, Fairchild was influential in persuading them to offer their customers plants that had a sporting chance of survival. Quite properly in 1730 his name headed the list of gardeners in the Society's *Catalogus Plantarum*, a comprehensive sales catalog focusing on exotic and domestic shrubs and trees sturdy enough, it was claimed, "to bear the cold of our climate in the open air."

Today Hoxton has become an area favored by urban professionals. The City Gardens disappeared long ago, and now only street signs bearing Fairchild's name show where his nursery used to trade. Fairchild himself died in 1729, leaving instructions that he be entombed in the humblest corner of the churchyard of St. Leonard's, Shoreditch, where the poor were usually buried. His bequest provided for the Worshipful Company of Gardeners to establish a tradition of giving a horticulturally themed sermon in the church each year. What an appropriate memorial to the gardener who united a love of science with the practice of his art!

For more information on environmental conditions in London at the time of Thomas Fairchild, see www.museum-london.org.uk.

Robert Prince's Long Island Nursery

America's first commercial nurseryman, Robert Prince, founded a horticultural dynasty which eventually became so influential that it even changed the trees on the streets of New York!

Of Robert Prince's early life virtually nothing is known, but in 1737 he created his famous Prince's Nursery on an 8-acre plot at Flushing Landing on Long Island, New York. Flushing was at that time little more than a small rural community and the business was built adjacent to Flushing Creek, allowing for the shipping of plants and trees via Flushing Bay. Robert was not only a skilled plantsman but also a shrewd entrepreneur who developed Prince's into a center of inter-colonial trade. As such it became a major importer of plants, and the leading exporter of American plants to Europe. In due course, when the plant hunters Meriwether Lewis and William Clark explored the north-west during the Jefferson Administration, many of the botanical treasures they found were sent back to Prince's Nursery.

While the business expanded to include shade and ornamental trees it remained renowned for its propagation and sale of fruit trees, especially the best and newest varieties of apple trees then being developed in the New World. There was an ever-increasing demand to produce varieties of the species brought over by the earlier settlers; the trees in particular had been so popular that they were at risk of being destroyed by overcutting. In 1771 Prince's published an advertisement famously offering for sale thirty-three different kinds of plum tree, forty-two types of pear, twenty-four varieties of apple, and twelve of nectarine. The work of the nursery was carried on after Robert's death by his sons Benjamin and William.

During the War for Independence, the British forces which occupied Long Island considered the nursery to be so special they put an armed guard around it to keep it safe. It was, happily, spared and the troops even allowed shipments for the Empress Josephine's rose-garden to travel unharmed. Even so, the nursery was cut off from most of the outside world and consequently business declined. It got so bad that at one point the nursery offered for sale grafted cherry trees, proposing that they could be used as barrel hoops. After Independence, the nursery's recovery was greatly helped when George Washington made a celebrated visit. Arriving by barge from New York, the inaugural President descended on the nursery and purchased several fruit trees.

Robert Prince's grandson, William Prince Jr., successfully used the kudos of the family name to establish a Linnean Botanic Garden. That, together with the Long Island nursery, which had now expanded to 111 acres, meant that his influence on American garden design was strong, and particularly in large estate gardens along the Hudson River. William's book, *A Short Treatise on Horticulture* (1828) enjoyed great popularity, as did his *Manual of Roses* (1846). By 1820 he had heard of a tree called "Heaven for sale," which he soon recommended for extensive use in landscaping American estates and parks. Such was the power of his enthusiasm that he even persuaded the growing metropolis of New York to use it as a street tree. Consequently, it began to replace the Linden oak and native maple, not only in New York but also in Baltimore, Philadelphia, and parts of Columbia.

Prince's nursery eventually shut down in 1865, by which time the Prince family business had achieved a lasting effect on the appearance of the American streetscape.

Further details at the Long Island history website:
www. lihistory.com.

Telford's Mail-Order Catalog

When a nationwide mail-coach service was established in Britain in the 1780s the Telfords of York, a dynasty of enterprising nurserymen, were in the fortunate position of being able to capitalize on it, having already established their nursery on the basis of selling by catalog. The founder of the dynasty, George Telford, leased some land in York in 1665 and seems to have started a nursery there. By the second half of the eighteenth century the same business was being run by his grandsons John Telford (1744–1830) and George Telford (1749–1834). By then Telford's was an old-established firm, predominantly concerned with retailing seedling trees and shrubs. In the 1750s and 1760s seeds formed a relatively insignificant proportion of their trade—indeed, records show they would often include seeds free of charge in an order.

The precedent of the Tradescants aside, the gardening catalog as a vehicle of trade began in earnest just after the Restoration of Charles II in 1660. A few nurseries in London and elsewhere had them printed, usually on a single page and without mentioning any prices. Telford's revolutionary catalog in the 1770s came fifty years after the effective start of a local press in York, indicating that by then the regional trade for horticultural items had become highly developed. The list that John and George Telford produced in 1775 was sensational because no one had ever before produced a catalog *with prices*.

Covering eighteen pages, their *Catalogue of Forest-Trees, Fruit-Trees, Ever-Green and Flowering-Shrubs* had the prices of stock trees and shrubs and other items marked throughout.

A
CATALOGUE

OF

FOREST-TREES,	**EVER-GREEN AND**
FRUIT-TREES,	**FLOWERING-SHRUBS,**

SOLD BY

JOHN and GEORGE TELFORD,

NURSERY-MEN AND SEEDS-MEN,

IN

TANNER-ROW, *YORK.*

Y O R K:

Printed by A. WARD, in Coney-Street, 1775.

Telfords's mail-order garden catalog, 1775.

CATALOGUE, &c.

	£.	s.	d.	
ASHES, one Year old Seedlings —		10		per thousand
Ditto transplanted, 1 Foot —	1	10		per thousand
Ditto —— 2 Feet —	2			per thousand
Ditto —— 3 Feet —		7	6	per hundred
Ditto				
Ash, Virginian Flowering —			6	each
Ashes, Mountain, 2 Feet and 3 —	1	5		per hundred
Ditto —— 4 Feet —	1	13	4	per hundred
Ditto —— 6 Feet —	2			per hundred
Ditto —— 8 Feet —	2	10		per hundred
Ash, Carolina —		1		each
Areatheophrasti, 2 Feet —	1	5		per hundred
Ditto —— 4 Feet —	1	13	4	per hundred
Ditto —— 6 Feet —	2			per hundred
Acerplatanoides, or Norway Maple, 3 Feet			6	each
BEECHES, one Year old Seedlings, 6 Inches		10		per thousand
Ditto, two Years old Seedlings —		15		per thousand
Ditto, transplantedd, 1 Foot —	2			per thousand
Ditto —— 1¼ Foot —		6		per hundred
Ditto —— 2 Feet —		7	6	per hundred
Ditto —— 3 Feet —		15		per hundred
Ditto —— 4 Feet —	1			per hundred
Beech, Purple —		5		each
Birches, transplanted, 2 Feet —		7	6	per hundred
Ditto —— 3 and 4 Feet —			3	each

A

The cost of one thousand oak seedlings, transplanted and 12 inches high, was just £1.50—enough to cover 10 acres! An asterisk was placed against plants that required the shelter of a glasshouse or a stove. Imitation being the sincerest form of flattery, Telford's brilliant priced-catalog idea was almost immediately copied by its main northern rivals, the Prefects of Pontefract.

The advent of the nationwide fast mail-coach system from 1784 onward enabled customers to receive Telford's catalog by post. Orders could be placed by mail, and the goods were dispatched by mail-coach at regular times, although quantity was necessarily limited. Large orders were placed through the catalog by the owners of Yorkshire's large country estates, such as Castle Howard and Duncombe Park. The new kinds of wooded landscapes being designed required a wide variety of trees, and it is clear that the Telfords played a very important part in producing the enormous stocks of young trees required for the great age of northern planting.

All this trade meant that Telford's of York boomed. Its premises grew to cover 19 acres on four plots, carrying a stock of some 350,000 trees, shrubs, and seedlings, plus countless seeds. The brothers became very wealthy, and by 1787 they had invested in the purchase of the manor of Eke, close to Beverley. Their nursery business had become the most important in the north but in 1816 they sold the enterprise as a going concern, not because they had reached retirement age, but in order to take up the even more profitable occupation of whaling.

 Visit the Post Office Heritage Museum, Freeling House, Phoenix Place, London WC1X 0DL. There is a useful website at the Smithsonian National Postal Museum: www.si.edu/postal.

CONTAINERS

Ramses III's Flowerpots

Ironically the humblest of garden artifacts began with the very grandest of innovators, Ramses III, the last of the great pharaohs, in approximately 1230 B.C. Or so it would seem. The earliest representation of a pot-grown plant is sometimes claimed to be evident on an altar decoration in the Maltese temple of Hagar Quim. Possibly the oldest representation of a plant container—if that is what it is—in the world, it appears to show a tall plant growing from a basin-shaped pot. That indicates, though does not prove, that there might have been some pot-growing activity prior to 2000 B.C.

Born in 1198 B.C., Ramses III was a reforming monarch who had the great advantage of indirectly succeeding (a few hundred years later) Ramses II who, by military conquest, had done much to enhance Egypt's security. Ramses III became a great benefactor and avid builder of temples and palaces, notably those cut into the rock at Abu Simbel in the Upper Nile area. Although the Egyptians had become avid gardeners, their plots were decidedly formal, and planting was done in a

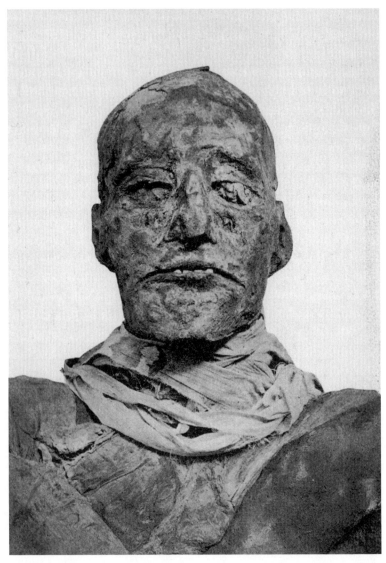

Ramses III ordered the creation of gardens with flowerpots for 514 Egyptian temples. (Mary Evans Picture Library)

systematic manner in straight lines. Ramses III established a staggering 514 semi-public gardens which, in the spirit of his relatively unstuffy approach, he fitted out with informal flowerpots. These earthenware containers could be planted with papyrus and multitudes of brilliant flowers and shrubs, then dotted along the walks. This initiative appears to have had the effect —unwittingly or not—of stimulating a much wider and informal use of flowerpots.

Geographical proximity makes it possible that the Ancient Greeks learned the practice of pot-gardening from Egypt in the sixth century B.C. For Greek women, the impetus was the need to grow produce—especially wheat, fennel, and lettuce—for the festival of Adonis. Seedlings were grown in a pot and allowed to wither, thus symbolizing the early death of Aphrodite's lover. Potted plants decorated with statues of Adonis were displayed on flat housetop roofs during this summer festival. Pots also had a curious signaling function.

A form of semaphore began in about 300 B.C. when the Greeks devised a means of arranging large vases on two low walls to spell out different letters of the alphabet.

The Romans also used a wide variety of pots on their roof gardens. Whereas the Greeks' motivation had been prompted by the desire to participate in a festival, the Romans—like the Egyptians in Ramses III's time—had a love of gardening. Eager to keep in touch with what they perceived to be their agricultural roots, urban-dwellers used pots to brighten their lives in the narrow, overcrowded and dingy streets. They liked informal groupings of windowsill pots from which would grow plants and vines on balconies and pergolas. At ground level much larger pots were used. Archaeologists at Pompeii have found sizable "planting holes" for terracotta pots embedded in the soil. These pots had holes in their sides, presumably to allow roots to grow out of the holes.

It was only in the 1600s that the practice of house plant cultivation really showed the slightest sign of spreading in Europe. In 1653 the agricultural guru Sir Hugh Platt produced *The Garden of Eden*, a reissue of the 1608 publication *Floraes Paradise*. Even so, decorative plant pots, known as jardinières, only became popular in England in the 1800s.

Ramses III's reign was not without incident. The first recorded strike in history occurred then, when workers building a necropolis protested against the government's failure to deliver grain rations. In his latter years one of his wives attempted to murder him. His death in 1166 B.C. was followed by centuries of weakness, leading to the Egyptian Empire being dominated by other countries.

 For further information on visiting Medinet Habu, the Temple of Ramses III, see www.touregypt.net.

Guen-ming's Bonsai

Most people think bonsai are a Japanese garden form, but in fact they originate from China, although there is some dispute as to precisely who instigated the practice of growing miniature gardens. An emperor in the Han Dynasty (206 B.C.–221 A.D.) created a miniature landscape in his courtyard, complete with valleys, caves, and trees, depicting his entire empire. It was said that anyone else caught in possession of even a part of a miniature landscape would be regarded as a threat to the empire—and executed. A more probable inventor was a high-ranking civil servant in the fourth century A.D., Ton Guen-ming. Also a famous poet, Guen-ming had retired to the country to grow chrysanthemums in pots—a move which reputedly led to his nurturing miniature trees. Known as *penjing*—literally, gardens in trays—this practice eventually formed the basis of bonsai.

Whether or not Guen-ming was the inventor of bonsai, in 1972 evidence was uncovered which proved that bonsai were known in 706 A.D. On opening the Tang Dynasty tomb of Prince Zhang Huai, archaeologists discovered wall paintings showing servants carrying what appeared to be trays of miniature plants. These paintings constitute the earliest known record of the art. At that time techniques were being developed to control the growth of miniature fruit trees in order to force peach and apricot trees to flower for the Chinese Lunar New Year. To achieve this, the trees needed to be pruned and wired into grotesque shapes to ensure they performed when required.

Ton Guen-ming's *penjing*, which the Japanese came to regard as the origin of the art of bonsai, did not arrive from the Chinese mainland until 1195 A.D., according to a Japanese scroll attributed to that period. Brought to Japan by Buddhist monks, for many years the art was known only to wealthy nobles, and it was regarded as a symbol of prestige and honor. Only after the Chinese invasion of Japan in the fourteenth century did the art of bonsai begin to be enjoyed by people at all levels of society.

The shaping of individual trees became more studied and refined, and small pieces of rock and sometimes moss were added to the container, thereby bringing the essence of the mountain and forest into the home or garden. During the Tokulgawa era, a period of peace and isolation which began in the seventeenth century, bonsai flourished with all the arts, but it was only at the end of this period that the term bonsai first came into use. The diminutive trees had been called *hachi-no-ki* for hundreds of years before an exclusive term was considered necessary.

A significant reason why bonsai is perceived by so many westerners as a Japanese, rather than a Chinese, invention is because it was the Japanese who deliberately introduced the art to Europe. In the late nineteenth century Japan's

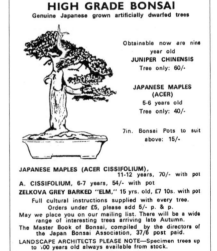

By the 1970s even general gardening magazines were running advertisements for bonsai products. (*The Gardeners' Chronicle*)

many centuries of isolation from the rest of the world came to an end, and Japanese businessmen attended international trade exhibitions in London and Vienna, and notably the Paris World Exhibition in 1900. These fostered interest in bonsai in the West and paved the way for the first major bonsai exhibition in London in 1909. Yet the public's initial reaction was not without hostility. In those early years Cassell's *Encyclopaedia of Gardening* described bonsai trees as "monstrosities." Some, perceiving that the trees looked tortured, expressed their displeasure at what seemed to be the cruel treatment of the trees by the bonsai masters. The Japanese had come to manipulate the trees with far more zeal than the art form's inventor, Guen-ming, could have intended. One wonders if he would have approved.

 Useful websites are Japanese Garden, www.jgarden.org; American Bonsai Society, www.absbonsai.org; and Bonsai Web, www.bonsai.com.

Nathaniel Ward's Wardian Case

In some respects Nathaniel Ward's accidental observation of matter growing in a bottle had all the momentous qualities of the discovery of penicillin. The contraption that Ward subsequently invented enabled two revolutions to occur: It provided a means by which plants could be grown indoors, and it allowed the long-distance transportation of plants.

The son of a successful doctor who practiced in the heavily polluted London docklands area, Nathaniel Bagshaw Ward was born on August 13, 1791. In his early years Nathaniel was strongly interested in a career at sea until, at the age of thirteen, he was sent on a vacation trip to Jamaica. To his father's relief the voyage changed Nathaniel's mind about seafaring, and his fascination with the island's coral reefs, tropical palms, and ferns established what was to become a lifelong enthusiasm for plants.

After completing his medical studies at the London Hospital, Nathaniel succeeded to his father's practice at Wellclose Square, Wapping. A pioneering enthusiast of "horticultural therapy," young Dr. Ward was a practical humanitarian who recognized the need for a means by which the urban poor—and particularly those who were sick, elderly, or mentally ill—could offset the grimness of their lives by means of contact with some form of greenery. An avid amateur naturalist, he strengthened himself to cope with surgeries by going on plant-collecting excursions to wooded and open spaces around London.

An early riser, each morning he studied plants and energetically tended his garden. For years he cherished an ambition to cover an old wall with rare ferns and mosses. To achieve it, he built up some rock work, then ingeniously placed a perforated pipe above it from which water trickled down on the mosses and ferns below. Despite his efforts, the heavily polluted smoke from the surrounding factories was so poisonous that the plants began to decline, then perished. Despondently wondering if he could ever grow anything worthwhile in such an environment he diverted his attention to raising butterflies.

During an experiment in the summer of 1829 to watch the hatching of a hawk moth, Ward buried a chrysalis in some moist leaf mold in a large wide-mouthed glass jar, which he then covered with a tin lid. Unbeknownst to him, a few seeds must have been lodged in the leaf mold. The insect attained its perfect form in about a month, and Ward was astonished to see two specks of vegetation apparently growing on the surface of the mold. Curious to observe the development of the fledgling plants in so confined a situation, he placed the bottle outside the north-facing window of his study—and to his amazement they flourished. Somehow, robustly thriving in the bottle were types of grass and a fern which he had by all other means been unable to grow in his garden. Also, and highly significantly, he noticed that during the heat of the day vapor regularly rose from the mold and the surface of the leaves, and condensed at night on the inside surface of the glass, before gradually trickling down into the soil. This recycling kept the mold at the same degree of humidity. In effect, Ward was witnessing the workings of the rain cycle in miniature, in his very own artificial micro-climate.

During the next three years or so Nathaniel beamed with pleasure whenever he glanced at the plants happily living in the glass bottle. No fresh water needed to be given, nor was the lid ever removed. The grass flowered and the fern produced three or four new fronds annually. Both plants ultimately perished only when rainwater found its way in through the rusted lid. Eager to find the limits of what his contraption could achieve, Ward was soon busily making other glass containers of various shapes and sizes. These "closely glazed cases," as he insisted on calling them, were nominally airtight except for two small, gauze-covered ventilator holes in the peak of each end. He repeated his initial experiment, with uniform success, upon some sixty types of fern, as well as various other plants (vascular as well as cellular), and he found plants that delighted in humid situations succeeded just as well as the ferns.

In March 1834 John Claudius Loudon, the gadget-loving editor of *The Gardener's Magazine*, received a tip-off that an eccentric doctor had discovered something special. He hastened over to Wellclose Square to see it for himself. That month he revealed to his readers that Ward had "the most extraordinary city garden we have ever beheld." Measuring 8 to 12 inches in width, some 30 inches in length, and 18 to 24 inches in height, the newfangled cases glinted in the sun everywhere—even on the roof and windowsills. There was a series of them along the tops of all the garden walls and adjoining offices, even on the slopes of lean-tos. Nonchalantly coming into bloom in a case in the drawing room was a magnificent specimen of *Melianthus major*. The largest and most audacious case was 10 feet high and covered 8.4 square feet. Already overgrown with fifty species of plants, Ward humorously dubbed it the "Tintern Abbey" case because it contained in its structure a scale model of Tintern Abbey's ruined west window.

A lesser-known element of Ward's glass jar findings, which he considered to be especially significant, was that the air of London was as well fitted to support plant life as the air of the country *once it was freed from soot*. The greatest advantage of air only being admitted into the case via imperceptible gaps in the lid was that by this means the air was filtered and impurities kept out. Thus, he reckoned, it was the soot in the capital's atmosphere, rather than the chemical composition of its air, which was so damaging to the environment. This was a particularly appealing notion to John Loudon, who was already pondering whether the quality of urban life might be improved by converting rooftops into conservatory garrets. On the basis of the Wardian case principle he now envisioned large greenhouses and housetop conservatories being constructed in the smoky air of London, and kept free from the impurities in the atmosphere by having all the air filtered through fine cloth. Similarly, it was believed, the purity of the air in living rooms might be increased by such filtration.

Other than Loudon the only person who knew the secret of this discovery was George Loddiges, proprietor of the famous Loddiges nursery gardens in nearby Hackney—one of Nathaniel's favorite haunts. The Ward, Loddiges, and Cooke families were interrelated: George Loddiges had married into the Cooke dynasty of maritime illustrators, and later so did Nathaniel's son Stephen. The Loddiges were pre-eminent sponsors of plant-finding expeditions, and from conversations with them Nathaniel became well aware that most plants dispatched from overseas failed to survive the voyage. All but a few would perish because of the drying winds and corrosive salt spray on deck, or from the lack of light below. Sensing that his invention might provide the protection required, in June 1833 Ward sent two glazed cases containing grasses and ferns to Sydney. Regardless of the range of climates weathered they arrived safely. So far, so good. The cases were then refilled with Australian plants and loaded aboard for the eight-month voyage home. Unwatered and untended on deck, they endured conditions varying from 20 degrees F and a 12-inch covering of snow on rounding Cape Horn, to 115 degrees F on the equator. On November 24, 1834, Nathaniel fretted nervously on the dockside as the sailing-ship came into view. Bounding up the gangplank the instant the ship's mooring lines were secured, he peered at the cases and whooped with delight. Not only had the cases functioned perfectly, but the beautiful coral ferns they contained were the first ever to be brought to Europe alive.

Wardian cases were not only used for conveying exotic plants about the world, they also made ideal indoor plant containers. (Hibberd, *Rustic Adornments*)

Wardian cases quickly became the standard method by which plant-finders moved their treasures around the world. Only a few refinements needed to be made to the basic design, such as the addition of louvered windows and roller blinds to protect the plants from the scorching sun. Loddiges, who had five hundred of them in operation by 1842, reported that whereas before they anticipated that nineteen out of twenty plants would perish, the trustworthy Wardian cases now meant they were surprised to lose more than one in twenty.

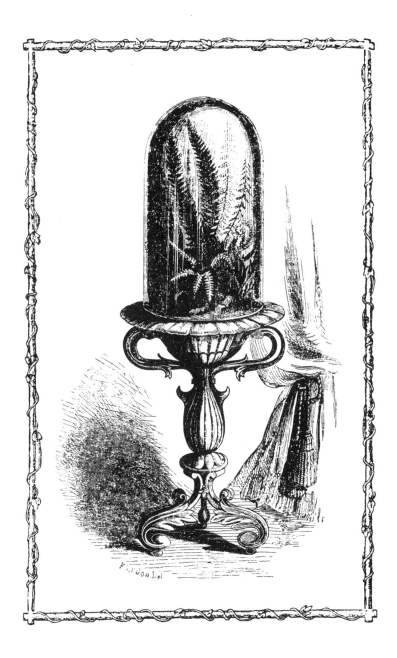

One of the earliest users was the plant-collector Robert Fortune who, disguised as a Chinese peasant, smuggled twenty thousand tea plant seedlings out of China. He took them from Shanghai to the Himalayas in Wardian cases, and thereby started India's tea industry. Such cases were also employed for the commercial introduction of the Cavendish banana from Chatsworth (where it had been successfully cultivated by Joseph Paxton) to the South Pacific. Irrespective of its humble origins, the Wardian case grew to have immense socio-economic importance as a unifying influence within the British Empire.

One of the most intriguing aspects of Ward's discovery was why it had come so late. After all, glass cloches had been in use for nearly two hundred years. Quite astonishingly, until Ward accidentally made his discovery, no one thought plants could live for more than a season under a glass dome. In fact, in 1825—four years before Ward—the unfortunate Scotsman Allan Alexander Maconochie, while independently experimenting with a miniature greenhouse and a large goldfish bowl, did discover the principle of closed glass containers. Maconochie did not make his findings public until 1839—whereas Ward had done so much earlier—and so, under the scientific "publish or perish" concept, Maconochie lost out on all claims to the invention. Being a gentleman (he was also a professor of geography), Maconochie chose not to press his claim, and moved to Australia where he became superintendent of a detention settlement on Norfolk Island.

The abolition of the Glass Tax in 1845, regarding which Ward was called to give evidence to a House of Commons Committee, and the introduction of the first sheet glass in 1848 should have meant that low-cost cases became affordable for millions of urban working-class people. For

266

them, Ward hoped that his wonder cases would be substitutes for poor views, or no views at all. Many of the cases were simply constructed of zinc on an iron framework. Often they were very ornate with decorative wrought-iron frames supported by carved stone bases. Despite his hopes, any sort of Wardian case still remained beyond the means of most of the poor, but they were taken up with gusto by the aspiring middle classes. In fashionable Victorian drawing rooms, which were so fume-filled that no plant could otherwise hope to survive, the cases were a novel focal-point. At the 1851 Great Exhibition, Ward displayed a case full of healthy plants which were still flourishing after eighteen years of not being watered! So versatile was the simple concept of the Wardian case that it spawned derivatives such as terrariums (which were partly enclosed), and even various forms of aquarium.

Personally quite unfazed by the momentousness of his discovery, Dr. Ward continued with his medical practice. He was made the Master of the Society of Apothecaries and a member of the Royal Society, although such honors meant little to him. A genuine philanthropist, he dithered about patenting his invention and consequently allowed himself to miss out on a fortune. He was more concerned that the closely glazed cases—he was too modest to call them Wardian cases—should be widely available to the poor, and to students for classroom demonstrations. He founded the Royal Microscopical Society in 1839, at a time when microscope soirées were regarded as a form of entertainment, and he often allowed his Wellclose Square house to be used as a gathering place for plant and microscope enthusiasts. After retiring from medical practice Ward moved to Clapham Rise, to a house appropriately named "The Ferns." In this less-polluted area he devoted himself to gardening. He died at St. Leonard's in Sussex in 1868, leaving a neatly arranged herbarium at The Ferns, which contained some twenty-five thousand specimens!

 A reproduction of an early Wardian case can be viewed at the National Maritime Museum, Greenwich, Park Row, London SE10; www.nmm.ac.uk.

🌱 Fisons's Gro-Bag

The Gro-Bag, a product that since 1973 has enabled millions of people to enjoy the simple pleasures of gardening, was not devised by an individual innovator but, uniquely, by a team of fairly obscure scientists. It was developed at Levington near Ipswich, on a research station established in the 1950s by Fisons Fertilizers Ltd. (a company founded by Joseph Fison in 1847). It was a logical extension of the Levington compost range, which was pioneered at Levington and launched in 1966. Levington compost was the first commercially successful peat-based growing medium. Lightweight, relatively cheap, reliable, and easy to use, the Levington compost range made pot and container gardening at home hassle-free and easy, and facilitated the production of the huge volumes of container-grown stock sold through the developing garden center industry.

Initially the growing bag was developed for professional tomato producers on the south coast and Channel Islands. Producers were finding that continual cultivation of tomatoes in the border soil of glasshouses was leading to a build-up of soil pests and diseases, severely affecting the yield. The commercial growing bag, introduced to the market in the late 1960s, offered a means of growing the crop in a pest- and disease-free environment.

Meanwhile, in 1965 a research team comprising three relatively unknown horticultural scientists, Peter Atkins, Derek Attenborough, and Brian Tree, at the Levington Research Station began to develop and hone the Gro-Bag, a product aimed at the general public. The contents were manufactured from fen peat, harvested from the Somerset levels, dried, and processed to a suitable particle size. Lime was then added to correct the pH and fertilizers mixed in to support initial plant growth. The formation of the compost seemed to be foolproof, but the scientists needed to know how inexperienced gardeners would take to raising plants in plastic bags. They ran consumer-testing trials in East London with people who lived in flats on a council estate. Each tenant was presented with one Gro-Bag, and for the next few months a camera crew filmed the transformation from an urban jungle into a plant haven. Some residents used the bags for growing herbs and salad vegetables, while others were successfully more adventurous with melons, squashes, runner beans, and even flowers.

Fisons launched the first Gro-Bag, for the amateur gardener in 1973. They were 5 feet long, 23 inches wide, and 9 inches deep, and each weighed approximately 35 pounds. The bags had special features such as loops on the ends which could be cut off and slipped around the center to hold everything together; there was also a yellow leaflet, supplied in a kangaroo-style pocket, which gave clear and concise instructions. Believing that a black bag would attract too much heat during the warm weather, Fisons launched the Gro-Bag with bright, almost garish, red and yellow

If you can't grow it in Fisons Gro-Bag you can't grow it.

Forget about not being born with green fingers. When you grow vegetables in a Fisons Gro-Bag,® you can't grow wrong.

A Gro-Bag can produce over 60lbs of plump tomatoes from just 3 plants. Think what that could save on your greengrocer's bill. It just goes to prove how superior the Fisons Gro-Bag still is.

*Gro-Bag is a registered trademark of Fisons Limited.

FISONS GRO-BAG

Fisons–the only and original Gro-Bag.

Gro-Bag advertisements became common in the late 1970s. (Courtesy of the Scotts Company)

packaging. This became a distinctive feature and was copied by most competitors. On several occasions attempts were made to introduce bags in more muted, garden-friendly, colors, but such versions never succeeded.

Fisons invested very heavily in advertising and promotion, and rapidly established a large consumer market in the UK. Although capable of growing a wide variety of crops, the growing bag was principally used by gardeners for tomato production. Properly watered and fed, three tomato plants in a Gro-Bag could yield up to 60 pounds of tomatoes. Growing bags quickly created a market worth millions of dollars— which represented almost one growing bag for every household with a garden in Britain! Despite efforts over subsequent years, using various techniques such as the introduction of mini-Gro-Bags, no one ever managed to establish similar markets in other countries—the amateur growing bag was a phenomenon unique to the UK.

The first commercially successful growing bag to be launched for the general public, the original Fisons's Gro-Bag (now owned and manufactured by the Scotts Company) remains the leading brand in the market. It is a success story in terms of consumer value for money, if not commercial profit. Launched in 1973, Gro-Bag is one of those brand names which has become a generic term. It began as a registered trademark but, like Hoover, is now so well used that it has won a place in the Oxford English Dictionary.

W. Atlee Burpee's Seeds

No company has had a greater impact on the American garden than Burpee. The Burpee company developed many plants and flowers that all gardeners today take for granted: Iceberg lettuce, yellow sweet corn, and the popular varieties of the marigold and zinnia. The arrival of the glossy Burpee catalog, bursting with scarlet tomatoes and clamoring annuals, is for gardeners an early sign of spring. Strange to think that the company's founder started out selling chickens.

Washington Atlee Burpee was born in 1858 in New Brunswick, New Jersey. His first passion was poultry: By age fourteen, the precocious Atlee was breeding chickens, geese, and turkeys and writing articles for poultry journals. W. Atlee

Burpee & Company was born in 1876 when eighteen-year-old Atlee dropped out of the University of Pennsylvania Medical School and launched a mail-order poultry and livestock business in Philadelphia with $1,000 loaned to him by his mother. In addition to animals, he offered seed so that his customers could raise their own feed. Before long Atlee noticed that he was receiving more orders for the seed than for the animals themselves, so in 1878 he began selling grain, flower, and vegetable seeds through the mail. Soon the seed sales dominated his business, and the animals were pushed to the back of the catalog.

Before Burpee entered the business, almost all seeds sold in America had been developed and produced in Europe. In 1888 Burpee, who was familiar with the work of geneticist Gregor Mendel, bought an eighteenth-century farm in Doylestown, Pennsylvania, named it Fordhook Farms, and there established America's first experimental farm for testing new varieties of vegetables and flowers and producing seeds. He spent summers touring Europe and the United States in search of superior vegetables and flowers, shipping thousands of specimens back to Fordhook for testing. The best plants were sold in his catalog or were crossed with stronger types to create offspring that would thrive in the United States. These methods produced some of Burpee's most famous varieties: Iceberg lettuce (1894), named for its crispness; the first jumbo bell pepper, Chinese Giant (1900); Golden Bantam, the first yellow sweet corn (1902); and the Fordhook Bush lima bean (1907), which is descended from a single plant that survived a cutworm infestation in 1890. In later years Burpee would broaden his testing grounds by establishing Floradale Farms in Lompoc, California, and Sunnybrook Farms near Swedesboro, New Jersey.

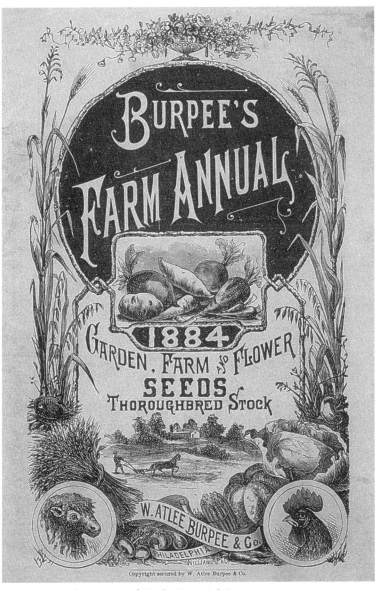

A Burpee seed packet. (Courtesy of W. Atlee Burpee & Co.)

A woodblock print of the old Burpee building. (Courtesy of W. Atlee Burpee & Co.)

Burpee's business model was no less innovative. Burpee took advantage of a major development in the U.S. postal system at the turn of the century. In 1863 the free delivery system was instituted in forty-nine of the country's largest cities, requiring the postman to bring letters directly to recipients' homes free of charge. Rural free delivery brought the same privilege to Americans living in rural areas beginning in 1896. Before then, the roughly thirty million Americans who lived in rural areas had to travel to the nearest post office, often miles away, to send and receive mail. The rural free delivery system put every gardener in America within easy reach of Burpee's colorful catalog.

Burpee quickly realized that advertising was the key to his business. In the early years the catalog was a mere 48 pages; by 1915 the catalog had swelled to two hundred pages and was mailed out in batches of one million. Burpee wrote most of the copy himself in a witty, engaging style, and in 1891 he began featuring engravings of plants based on photographs rather than hand-drawn illustrations. Alongside plant descriptions were stories of Burpee's travels around the world in search of exotic plants, and testimonials from satisfied customers. He created an advertising department within the company and offered cash prizes for the best advertising slogans, an approach which in 1890 yielded the line "Burpee's Seeds Grow," one of the great corporate slogans.

Despite a knack for marketing, Burpee did not misrepresent his seeds. The story goes that in 1914 Burpee was appalled to find on the cover of his catalog an image of a Matchless tomato that was larger than any he had ever seen in the field. Burpee printed an apology to his readers—and then began receiving letters from customers who swore the Matchless really did get that big. From the very start, Burpee unconditionally guaranteed his products for a full year after purchase. By the time of his death in 1915, Atlee's company had grown to become the largest mail-order seed company in the world.

Workers pack seeds to ship from Philadelphia. (Courtesy of W. Atlee Burpee & Co.)

When Atlee died his twenty-two-year-old son David dropped out of Cornell University to run the family business. Under David's stewardship the Burpee company focused on hybrids and flowers. Beginning in the 1930s, Burpee breeders used various methods, from classic Mendelian crosses to X-ray and chemical treatments of seeds, to produce hybrid vegetables and flowers with desirable or novel traits. Major breakthroughs included the Red and Gold marigold, the first hybrid flower from seed; new forms of petunias and nasturtiums; the Ambrosia cantaloupe; and in 1949 the Big Boy tomato.

Like his father, David had a flair for publicity. In 1954 he offered a $10,000 prize to the first person who could bring him the seeds of a coveted freak of nature, a white marigold. After twenty years and some $250,000 in research and evaluation expenses, Alice Vonk of Iowa was named the winner in 1975. During the 1960s—at a time when he was selling large quantities of marigold seed—David launched a campaign to make the marigold America's national flower. Despite the vigorous support of the Republican Senate majority leader, Everett Dirksen of Illinois, the bill failed, although it may have succeeded as a marketing stunt.

In 1970 David Burpee sold the company to General Foods, which in turn sold it to ITT in 1979. In 1991 the Burpee company was acquired by George J. Ball, Inc., a horticultural business. In recent years, as American gardeners have less free time and keep smaller plots, the Burpee company has branched out into selling live plants through the mail, demonstrating that a company that began with innovation has continued to evolve. In 1999 George Ball Jr., Burpee's president and CEO and himself a third-generation seedsman, purchased Fordhook Farms from W. Atlee Burpee's descendants. The facility has been re-established as a trial ground and today tests Burpee varieties before they are sold through the catalog—just as it did in the 1890s.

 For more on the Burpee company, see www.burpee.com.

Sources

CHAPTER 1. TOOLS OF THE TRADE
Chuko Liang's Wheelbarrow
Lewis, M., "The Origins of the Wheelbarrow," *Technology and Culture* (July 1994)
Matthies, Andrea, "The Medieval Wheelbarrow," *Technology and Culture* (January 1991)
Needham, Noel Joseph, *Science and Civilisation in China*, vol. 4, part 2 (Cambridge University Press, 1965)

Bertrand de Moleville's Secateurs
Allan, Mea, *William Robinson 1838–1935* (Faber & Faber, 1982)
Elliot, Brent, "Shear diversity," *The Garden* (March 1996)
Sanecki, Kay, *Old Garden Tools* (Shire, 1979)

George Acland's Jute Twine
Woodhouse, T. and P. Kilgour, *The Jute Industry* (Pitman, 1921)
British Jute Trade Research Association, *Jute* (1962).

William Barron's Tree-Mover
Barron, William, *The British Winter Garden* (1852)
The Gardeners' Chronicle (April 18 and 25, 1891)
Hadfield, Miles, "Tree Moving," *The Gardeners' Chronicle* (February 1, 1967)
Hinde, Thomas, *Capability Brown: The Story of a Master Gardener* (Hutchinson, 1968)
Robinson, William, *Gleanings from French Gardens*, 2nd edition (Frederick Warne, 1869)
Steuart, Henry, *The Planter's Guide* (John Murray, 1827)

Wilks, J., *Trees of the British Isles in History & Legend* (Frederick Muller, 1972)

Carl Nyberg's Flame-Gun
Abbis, H.W., "The Technique of Flame-Gunning," *The Gardener's Chronicle* (June 18, 1955)
De Bono, Edward, *Eureka!* (Thames & Hudson, 1974)

CHAPTER 2. PLANT FINDERS
Pietro Matthiolus's Tomato
DuBose, Fred, *The Total Tomato* (Harper Colophon, 1985)
Ferri, Sara, *Pietro Andrea Mattioli* (Quattromme, 1997)
Green, Edward Lee, *Landmarks of Botanical History* (Stanford University Press, Part II, 1983)
Heiser, Charles, *Nightshades: The Paradoxical Plants* (Freeman & Co., 1969)
Shewell-Cooper, W.E., *The Book of the Tomato* (John Gifford, 1949)

Mark Catesby's New World Depictions
The Connoisseur "A Review of Bird Prints" (1948)
Catesby, Mark, *Hortus Europae Americanus* (1747)
Catesby, Mark, *The Natural History of Carolina, Florida and the Bahama Islands* (1731)
Frick, George and Raymond Stearns, *Mark Catesby: The Colonial Audubon* (University of Illinois Press, 1961)
Miller, William, "Mark Catesby, Eighteenth Century Naturalist," *Journal of the New York Botanical Garden* (January 1949)

Silver, Bruce, "Mark Catesby and the Colonial American Wilderness," *Ex Libris* (Spring/Summer 1981)

Willson, E.J., *West London Nursery Gardens* (Fulham and Hammersmith Historical Society, 1982)

John Bartram's Plant Boxes

Culross Peattie, Donald, *Green Laurels: The Lives and Achievements of the Great Naturalists* (Simon & Schuster, 1936)

Dunning, Brian, "American who Changed English Gardens," *Country Life* (September 8, 1966)

The Philadelphia Inquirer, "John Bartram Garden Still Flourishing" (April 7, 1963)

Povey, D., "Garden of a King's Botanist," *Country Life* (March 22, 1956)

Read Cheston, Emily, *John Bartram 1699–1777* (John Bartram Association, 1953)

Stewart, Robert G., "A Portrait of John Bartram Identified," *The Garden Journal* (January–February 1967)

Nathaniel Wallich's Rhododendron

Buchan, Ursula and Nigel Colborn, *The Classic Horticulturalist* (Cassell, 1987)

Coats, Alice, *The Quest for Plants* (Studio Vista, 1969)

Davidson, William, *Exotic Indoor Plants* (Macdonald, 1984)

Rose, Dorothy, "Dr. Nathaniel Wallich," *The Garden* (May 1981)

Santapau, H., "The Indian Botanic Garden," *Bulletin of the Botanical Survey of India*, vol. 7 (1965)

"Chinese" Wilson's Regal Lily

Allan, Mea, *Plants that Changed Our Gardens* (David & Charles, 1974)

Briggs, Roy, *"Chinese" Wilson* (HMSO, 1993)

Heriz-Smith, Shirley, "Choicest Chinese Take-Away," *Country Life* (May 18, 1989)

Hiller, Harold, "E.H. Wilson," *The Garden* (October 1976)

Lyte, Charles, *The Plant Hunters* (Orbis Publishing, 1983)

Musgrave, Toby, Chris Gardner, and Will Musgrave, *The Plant Hunters* (Ward Lock, 1998)

Rehder, Alfred, "Ernest Henry Wilson," *Journal of the Arnold Arboretum*, vol. XI (1930)

CHAPTER 3. WATER FEATURES

Archimedes's Waterscrew

Bendick, Jeanne, *Archimedes and the Door of Science* (Chatto & Windus, 1964)

Desmond, Ray, *Kew: The History of the Royal Botanic Gardens* (Harvill, 1995)

MacDougall, Elisabeth B. and Naomi Miller, *Fons Sapientiae: Garden Fountains in Illustrated Books* (Dumbarton Oaks, 1977)

Thomas Hyll's Sprinkler

Johnson, George W., *A History of English Gardening* (Baldwin & Cradock, 1829)

Ignace Dubus-Bonnel's Fiberglass Pond

Ledbetter, Gordon, *Water Gardens* (Alpha Books, 1979)

Newman, Roger, "Pools to Last a Lifetime," *Gardeners' Chronicle* (March 10, 1972)

Robinson, Peter, *Water Gardening* (Dorling Kindersley, 1997)

Russell, Stanley, *The Stapeley Book of Water Gardens* (David & Charles, 1985)

Henry Bewley's Gutta Percha Hose

Brant, William T., *India Rubber, Gutta-Percha, and Balata* (Sampson Low, 1900)

James, John, *The Theory and Practice of Gardening* (1712); reprint by Gregg International, 1969

Schidrowitz, P. and T. Dawson, *History of the Rubber Industry* (Heffer, 1952)

CHAPTER 4. GROWING EXOTICS
Salomon de Caus and his Orangery
Loudon, John Claudius, *An Encyclopaedia of Gardening* (1822)
Strong, Roy, *The Renaissance Garden in England* (Thames & Hudson, 1979)

Henry Telende's Pineapple
Bradley, Richard, *Treatise on Ancient Husbandry & Gardening* (1726)
Collins, J.L., *The Pineapple* (Leonard Hill Books, 1960)
Huxley, Anthony, *An Illustrated History of Gardening* (Paddington, 1978)
Levins, Hoag, *Symbolism of the Pineapple* (2000)

Chabannes's Greenhouse Central Heating
Beeton, Isabella, *The Book of Garden Management* (Ward Lock, 1872)
Gardener's Magazine (March 1828)
Hix, John, *The Glass House* (Phaidon, 1974)
Kolmaier, Georg and Bana von Sartory, *Houses of Glass* (MIT Press, 1986)
Lemmon, Kenneth, *The Covered Garden* (Museum Press, 1962)
Loudon, John Claudius, *An Encyclopaedia of Gardening* (1822)
M'Intosh, Charles, *The Book of the Garden* (1853)
Saint, Andrew et al., *A History of the Royal Opera House* (Royal Opera House Publications, 1982)
Van de Muijzenberg, E., *A History of Greenhouses* (Institute for Agricultural Studies, Mansholtlan, 1980)

George Stephenson's Cucumber
Davies, Hunter, *George Stephenson* (Weidenfeld & Nicolson, 1975)
Parsionage, W.R., *A Biography of George Stephenson* (Institute of Mechanical Engineers, 1937)
Smiles, Samuel, *The Life of George Stephenson* (John Murray, 1881)

Ephraim Wales Bull's Concord Grape
Collins, Paul, *Banvard's Folly: Thirteen Tales of Renowned Obscurity, Famous Anonymity, and Rotten Luck* (Picador, 2001)

CHAPTER 5. LAWNS
Pliny the Younger's Lawn
Castell, Robert, *Villas of the Ancients* (1728)
Cuthbertson, Yvonne, *Women Gardeners: A History* (Arden Press, 1998)
Du Prey, P., *The Villas of Pliny* (1994)
Loudon, John, *An Encyclopaedia of Gardening* (1822)
Scott Jenkins, Virginia, *The Lawn: A History of an American Obsession* (Smithsonian Institutional Press, 1994)
Teyssot, George (ed.), *The American Lawn* (Princetown Architectural Press, 1999)

John Jaques II's Croquet
Gill, A.E., *Croquet: The Complete Guide* (Heinemann, 1988)
Pritchard, David, *The History of Croquet* (Cassell, 1981)
Jaques, John, *Croquet, the Laws and Regulations of the Game* (Jaques, 1864)

Charles Cathcart the Moss-Gatherer
Elliot, I.M.Z., *A Short History of Surgical Dressing* (London Pharmaceutical Press, 1964)
Power, Sir d'Archy, *Lives of the Fellows of the Royal College of Surgeons* (1953)
Schenk, G., *Moss Gardening* (1967)

Karl Dahlman's Flymo
Stewart, P.S., *The Economics of Invention and Innovation with a Case Study of the Development of the Hovercraft* (1975)

Frederick Law Olmsted's Great Lawn
Stevenson, Elizabeth, *Park Maker: A Life of Frederick Law Olmsted* (Transaction Publishers, 2000)

CHAPTER 6. SUPPORTS, CLIMBERS, AND HEDGES

Cicero's Trellis
Farrar, Linda, *Ancient Roman Gardens* (Sutton, 1998)
Garnock, Jamie, *Trellis* (Thames & Hudson, 1991)
Strong, Roy, *The Garden Trellis* (Pavilion, 1991)

Christopher Columbus and the Wigwam Frame
Dunster, Henry P., *The Discoveries of Columbus and of the English in America* (James Blackwell, nd)
Fernandez-Armesto, Felipe, *Columbus* (OUP, 1991)
Frampton, John, *Joyful News Out of the New-Found World* (1580)

William Chambers's Bamboo
Bald, R.C., "Sir William Chambers and the Chinese Garden," *Journal of History of Ideas*, no. 2 (1950)
Blunt, Wilfred, *In for a Penny* (Hamish Hamilton, 1978)
Chambers, William, *Designs for Chinese Buildings at Kew* (1763)
Chambers, William, *Dissertation on Oriental Gardening* (1772)
Farrelly, David, *The Story of Bamboo* (Sierra Club, 1984)
Freeman-Mitford, Algernon, *The Bamboo Garden* (Macmillan, 1896)
Hardwick, Thomas, *A Memoir of the Life of Sir William Chambers* (1825)
High Lawson, Alexander, *Bamboos* (Faber & Faber, 1968)
Pevsner, Nikolaus, "The Other Chambers," *Architectural Review* (1947)
Sirén, Osvald, *China and Gardens of Europe of the Eighteenth Century* (Ronald Press, 1950)

Christopher Leyland's Monster Hedge Tree
Bean, W.J., *Trees and Shrubs* (John Murray)
Jackson, A. and W. Dallimore, "A New Hybrid Conifer," *Kew Bulletin*, no. 3 (1926)
Ovens, H., W. Blight and A. Mitchell, "The Clones of Leyland Cypress," *Quarterly Journal of Forestry* (January 1964)
Slinn, Joy, *A Souvenir History of Haggerston Castle* (Brownlamb Books, 1995)

CHAPTER 7. FERTILIZERS AND PEST CONTROLS

Columella's Soil Test
Glacken, Clarence J., *Traces on the Rhodian Shore* (University of California Press, 1973)
Janson, H. Frederic, *Pomona's Harvest: An Illustrated Chronicle of Antiquarian Fruit Literature* (Timber Press, 1996)

Leonard Mascall, Grub Catcher
The Cottage Gardener (July–August 1851)
Journal of Horticulture and Cottage Gardener (January 28, 1875)
Loudon, John, *An Encyclopaedia of Agriculture* (Longman, 1825)
McDonald, Donald, *Agricultural Writers* (Horace Cox, 1908)
Mascall, Leonard, *The Country Man's Recreation or the Art of Planting, Grafting and Gardening* (1640)
Mascall, Leonard, *On the Government of Cattell* (1596)
Thompson, Robert, *The Gardener's Assistant* (Gresham, 1905)

Robert Sharrock's Mole Trap
Bucjacki, Stefan, *Best Garden Doctor* (Hamlyn, 1997)
Bucjacki, Stefan, *Garden Warfare* (Souvenir Press, 1988)
Carnegie, William, "How to Trap and Snare," *Shooting Times* (nd)
Godfrey, G. and P. Crowcroft, *The Life of the Mole* (Museum Press, 1960)
Gorman, Martyn L. and R. David Stone, *The Natural History of Moles* (Christopher Helm, 1990)

Olkowski, William, Sheila Daas, and Helga Olkowski, *Common-Sense Pest Control* (Taunton Press, 1991)

Sharrock, Robert, *An Improvement to the Art of Gardening* (Nathanael Sackett, 1694)

William Forsyth's Tree Plaster
Chelsea Physic Garden, *William Forsyth 1737–1804* (1999)

Forsyth, William, *Observations on the Diseases, Defects and Injuries of Fruit and Forest Trees* (1791)

Forsyth, William, *Treatise on the Culture and Management of Fruit-trees* (1802)

Meynell, Guy, "The Personal Issue Underlying T.A. Knight's Controversy with William Forsyth," *Journal of the Society for the Bibliography of Natural History* (November 1979)

Minter, Sue, *The Apothecaries' Garden* (Sutton, 2000)

Simmonds, A., *A Horticultural Who was Who* (RHS, 1948)

John Bennet Lawes's Superphosphate
Dyke, G.V., *John Lawes of Rothamsted* (Hoos Press, 1993)

Rothamsted Experimental Station, *Rothamsted: One Hundred and Fifty Years of Agricultural Research* (1993)

George Washington Carver's Peanut
Kremer, Gary, *George Washington Carver in His Own Words* (University of Missouri, 1987)

Manber, David, *The Wizard of Tuskegee: The Life of George Washington Carver* (Crowell-Collier Press, 1967)

CHAPTER 8. GARDEN WRITINGS
Richard Bradley's History of Succulent Plants
Bradley, Richard, *The Country Housewife & Lady's Director*, reprinted with an introduction by Caroline Davidson (Prospect Books, 1980)

Roberts, W., "Richard Bradley, Pioneer Garden Journalist," *Journal of the RHS* (April 1939)

Rowley, Gordon, *A History of Succulent Plants* (Strawberry Press, 1997)

Carl Linnaeus and His Plant Classification
Landell, Nils-Erik, *The Gardener Linne* (Carlsson, 1997)

Overy, Angela, *Sex in Your Garden* (Fulcrum, 2000)

Stearn, William and Gavin Bridson, *Carl Linnaeus* (Linnean Society of London, 1978)

Jane Loudon's Gardening for Ladies
Gloag, John, *Mr. Loudon's England* (Oriel Press, 1970)

Hadfield, Miles, *Pioneers in Gardening* (Routledge & Kegan Paul, 1955)

Howe, Bea, *The Lady with Green Fingers* (Country Life, 1961)

Loudon, Jane, *Gardening for Ladies* (John Murray, 1840)

Taylor, Geoffrey, *Some Nineteenth-Century Gardeners* (Skeffington, 1951)

James Shirley Hibberd's Rustic Adornments
Gorer, Richard, "The Victorian who said it all," *Country Life* (March 13, 1980)

Hibberd, Shirley, *Rustic Adornments for Homes of Good Taste*, with an introduction by John Sales (Century Hutchinson, 1987)

Wilkinson, Anne, "The Preternatural Gardener: The Life of James Shirley Hibberd," *Garden History* (Winter 1998)

William Robinson's English Flower Garden
Allan, Mea, *William Robinson: Father of the English Flower Garden* (Faber & Faber, 1982)

Robinson, William, *The English Flower Garden*, sixteenth edition revised by Roy Hay (John Murray, 1956)

Robert B. Thomas and The Old Farmer's Almanac
The Old Farmer's Almanac (Yankee Publishing Inc., Dublin, New Hampshire)

CHAPTER 9. GARDENING MOVEMENTS
Theophrastus's Roses
Hadfield, Miles, Pioneers in Gardening (Routledge, 1955)
Hawks, Ellison, Pioneers of Plant Study (Sheldon Press, 1928)
Isley, Duane, One Hundred and One Botanists (Iowa State University, 1994)

Alexandre Dumas and the Camellia Craze
Coats, Peter, Flowers in History (Weidenfeld & Nicolson, 1970)
Hemmings, Frederic, The King of Romance (Hamish Hamilton, 1979)
Stirling, Macoboy, The Colour Dictionary of Camellias (2000)

Titus Salt's Garden Plots
Bullock, Rob and Gillie Gould, The Allotment Book (1988)
Crouch, D. and C. Ward, The Allotment (1988)
Hole, James, The Homes of the Working Classes (1966)
Reynolds, Jack, The Great Paternalist (London & Bradford, 1983)
Suddards, R. (ed.), Titus of Salts (1976)

Mary Montagu's Language of Flowers
Ingrams, John Henry, Flora Symbolica (1868)
Montagu, Lady Mary Wortley, Letters from Turkey (1763)
Robinson, Fanny, The Country Flowers of a Victorian Lady, ed. Gill Saunders (Apollo, 1999)
Ross, Stephanie, What Gardens Mean (University of Chicago Press, 1998)
Seaton, Beverly, The Language of Flowers (University of Virginia, 1995)

Eve Balfour, the Soil Association Guru
Balfour, Eve, The Living Soil and the Haughley Experiment (1943)
Carson, Rachel, Silent Spring (1963)
Conford, Philip, "Breaking New Ground," Living Earth (July 1996)
Dudley, Nigel, The Soil Association Handbook (Macdonald, 1991)
Gear, Alan, Jackie Gear, Pauline Pears, Bob Sherman, and Sue Stickland, Organic Gardening (Thorsons, 1991)
Langman, Mary, "As it was in the beginning," Living Earth (July 1996)

CHAPTER 10. NURSERIES
John Gerard the Herbalist
Buchan, Ursula and Nigel Colburn, The Classic Horticulturalist (Cassell, 1987)
Gerard, John, The Herball, or Generall Historie of Plantes (1597)
Hadfield, Miles, Pioneers in Gardening (Routledge, 1955)

John Tradescant's "Ark"
Allan, Mea, The Tradescants (1964)
Harvey, John, Early Nurserymen (Phillimore, 1974)
Leith-Ross, Prudence, The John Tradescants (Peter Owen, 1984)
Leith-Ross, Prudence, The Story of the Tradescants (Museum of Garden History, 1985)
Sturdy, David, "The Tradescants at Lambeth," Journal of Garden History (January 1992)

Thomas Fairchild's "City Gardens"
Barnes, Melvyn, Root & Branch (The Worshipful Company of Gardeners of London, 1994)
Evelyn, John, Fumifugium: or the Inconvenience of the Air and Smoke of London (1661)
Fairchild, Thomas, The City Gardener (1722)
Galinou, Mireille (ed.), The Glorious History of the Capital's Parks and Gardens (Anaya, 1990)

Leapman, Michael, *The Ingenious Mr. Fairchild* (Headline, 2000)

Lievre, Audrey Le, "Hoxton's Horticulturist," *Country Life* (November 3, 1988)

Robert Prince's Long Island Nursery

Hedrick, U.P., *A History of Horticulture in America to 1860* (OUP, 1950)

Prince, William, *A Short Treatise on Horticulture* (1828)

Punch, Walter, *Keeping Eden: A History of Gardening in America* (Bullfinch Press, 1992)

Shaver Haughton, Clare, *Green Innovators: The Plants that Transformed America* (Harcourt Brace Jovanovich, 1978)

Telford's Mail-Order Catalog

Harvey, John, *Early Gardening Catalogues* (Phillimore, 1972)

Harvey, John, *Early Horticultural Catalogues* (University of Bath, 1973)

Harvey, John, "The Family of Telford, Nurserymen of York," *The Yorkshire Archaeological Journal*, vol. XLII (1969)

CHAPTER 11. CONTAINERS

Ramses III's Flowerpots

Berrall, Julia, *The Garden* (1966)

Brace, Josh, *The Culture of Fruit Trees in Pots* (John Murray, 1904)

Hart, Colin, "Curiosities of Yesteryear," *Gardeners' Chronicle* (December 20, 1967)

Reader's Digest, *The Origins of Everyday Things* (1998)

Guen-ming's Bonsai

Keane, Marc P., *Japanese Garden Design* (Charles E. Tuttle, 1996)

Page, Christine, "The Origins of Bonsai," *Gardeners' Chronicle* (July 5, 1968)

Page, Christine, "Bonsai Culture," *Gardeners' Chronicle* (July 5, 1968)

Nathaniel Ward's Wardian Case

Lievre, Audrey Le, "Gardens among the Chimneypots," *Country Life* (March 1989)

Ward, Nathaniel, *On the Growth of Plants in Closely Glazed Cases* (John Vorst, 1852)

Westland, Pamela, *Terrariums* (Apple Press, 1993)

Fisons's Gro-Bag

Gardeners' Chronicle, "Alexpeat" (May 12, 1992)

W. Atlee Burpee's Seeds

Kraft, Ken, *Garden to Order* (Doubleday, 1963)

Useful websites

American Horticultural Society—
www.ahs.org

Biography—
www.biography.com

Garden Com—
www.garden.com

Gardenlinks—
www.gardenlinks.ndo.co.uk

Garden visits—
www.gardenvisit.com

Garden Web—
www.gardenweb.com

Garden World—
www.gardenworld.co.uk

History—
www.history

History of Horticulture—
www.hcs.ohio-state.edu/hor/history

Horticulture—
www.horticulture

I Love Gardens—
www.ilovegardens.com

Royal Horticultural Society—
www.rhs.org.uk

Vegetable Patch—
www.thevegetablepatch.com

Virtual Garden—
www.vg.com

Index